Developments in Atmospheric Science, 22

EVAPORATION, EVAPOTRANSPIRATION AND CLIMATIC DATA

Further titles in this series

1. F. VERNIANI (Editor)
Structure and Dynamics of the Upper Atmosphere

2. E. E. GOSSARD and W. H. HOOKE
Waves in the Atmosphere

3. L. P. SMITH
Methods in Agricultural Meteorology

4. O. ESSENWANGER
Applied Statistics in Atmospheric Science

5. G. W. PALTRIDGE and C.M.R. PLATT
Radiative Processes in Meteorology and Climatology

6. P. SCHWERDTFEGER
Physical Principles of Micro-Meteorological Measurements

7. S. TWOMEY
Atmospheric Aerosols

8. E. FUKUI (Editor)
The Climate of Japan

9. A. L. FYMAT and V. E. ZUEV (Editors)
Remote Sensing of the Atmosphere: Inversion Methods and Applications

10. W. BACH, J. PANKRATH and W. KELLOGG (Editors)
Man's Impact on Climate

11. A. LONGHETTO (Editor)
Atmospheric Planetary Boundary Layer Physics

12. C. MAGONO
Thunderstorms

13. S. IKEDA, E. SUZUKI, E. UCHIDA and M. M. YOSHINO (Editors)
Statistical Climatology

14. E. E. GOSSARD and R. G. STRAUCH
Radar Observation of Clear Air and Clouds

15. W. SCHWERDTFEGER
Weather and Climate of the Antarctic

16. A. L. BERGER and C. NICOLIS (Editors)
New Perspectives in Climate Modelling

17. R. W. PREISENDORFER and C. D. MOBLEY
Principal Component Analysis in Meteorology and Oceanography

18. D. MARTYN
Climates of the World

19. M. E. SCHLESINGER (Editor)
Greenhouse-Gas-Induced Climatic Change: A Critical Appraisal of Simulations and Oberservations

20. R. REITER
Phenomena in Atmospheric and Environmental Electricity

21. S.G. DOBROVOLSKI
Global Climatic Changes in Water and Heat Transfer-Accumulation Processes

Developments in Atmospheric Science, 22

EVAPORATION, EVAPOTRANSPIRATION AND CLIMATIC DATA

R. BURMAN
L.O. POCHOP

University of Wyoming
Department of Civil and Architectural Engineering
Colllege of Engineering
P.O. Box 3295
Laramie, Wyoming 82071-3295
U.S.A.

ELSEVIER
Amsterdam – Lausanne – New York – Oxford – Shannon – Tokyo 1994

ELSEVIER SCIENCE B.V.
Sara Burgerhartstraat 25
P.O. Box 211, 1000 AE Amsterdam, The Netherlands

ISBN 0-444-81940-1

This book is printed on acid-free paper.

Printed in The Netherlands

Table of Contents

Table of Figures

Table of Tables

LIST OF SYMBOLS

symbol	description	units
	English letters	
A_w	fraction of PESW	% or decimal
C	fractional cloud cover	dimensionless
C_p	specific heat at constant pressure	$kJ\ kg^{-1}\ K^{-1}$
C_s	stage 2 evaporation rate	$mm\ day^{-1}$
CU	consumptive use	$mm\ day^{-1}$
D	soil water depletion	mm
D_r	soil water depletion ratio	% or decimal
D	Julian or calendar day	days
E	evaporation from a water surface	$mm\ day^{-1}$
E_a	aerodynamic wind function	$mm\ cm^{-2}\ day^{-1}$
EL	lake evaporation	$mm\ day^{-1}$
E_p	pan evaporation	$mm\ day^{-1}$
ESL	shallow lake evaporation	$mm\ day^{-1}$
ET	evapotranspiration rate	$mm\ day^{-1}$
ET_a	actual evapotranspiration	$mm\ day^{-1}$
ET_m	maximum evapotranspiration	$mm\ day^{-1}$

ET_a	reference crop evapotranspiration	mm day $^{-1}$
ET_{ra}	reference crop ET, alfalfa	mm day $^{-1}$
ET_{rg}	reference crop ET, grass	mm day $^{-1}$
G	soil heat flux	MJ m^{-2} day^{-1}
H	time	hours
H	humidity	% or decimal
H	energy	m
K_c	crop coefficient, general	dimensionless
K_{cc}	crop coefficient, alfalfa based	dimensionless
L	long-wave radiation	MJ m^{-2} day^{-1}
L	latitude	degrees
LAI	leaf area index	dimensionless
M	longitude	degrees
MAD	management allowed deficiency	%
N	number of days in a month	
N	possible hours of sunshine	hours
P	pressure	kPa
P	vegetative production	kg m^{-2}
P	fraction of time	decimal or %
P_e	effective precipitation	mm
PRN	slope of production function	dimensionless

Q1	energy inflow to a lake	mm cm^{-2} day^{-1}
Q2	energy outflow from a lake	mm cm^{-2} day^{-1}
R	T_{max} - T_{min} in Linacre pan E method	°C
R	specific gas constant	kJ kg^{-1} K^{-1}
R	average R in Linacre pan E method	°C
R	long or short wave radiation	MJ m^{-2} day^{-1}
R_a	extraterrestrial solar radiation	MJ m^{-2} day^{-1}
R_b	net outgoing long wave radiation	MJ m^{-2} day^{-1}
Ri	Richardson number	dimensionless
R_n	net radiation	MJ m^{-2} day^{-1}
R_{nl}	net long wave solar radiation	MJ m^{-2} day^{-1}
R_s	solar radiation, diffuse & direct	MJ m^{-2} day^{-1}
R_{so}	clear day solar radiation	MJ m^{-2} day^{-1}
RH	relative humidity	% or decimal
S	sunshine ratio	dimensionless
S	shortwave radiation	MJ m^{-2} day^{-1}
S1	energy storage in a lake start of a day	mm cm^{-2} day^{-1}
S2	energy storage in a lake end of a day	mm cm^{-2} day^{-1}
SC	solar constant	MJ m^{-2} day^{-1}
T	temperature	°C or K

T	transpiration	mm day^{-1}
T_a	mean daily temperature	°C or K
T_k	temperature $T_K = T + 273.15$	K Kelvins
T_m	temperature function, Linacre method	°C
U	limit of stage 1 soil evaporation	mm
VPD	vapor pressure deficit	kPa
WCF	winter cover factor	dimensionless
WUE	water use efficiency	kg m_{-3}
Z	type of soil dryness curve	
c	Penman eq. correction factor	dimensionless
d	zero plane displacement	m
d	distance earth to sun	not specified
$\overline{d}$	average distance earth to sun	not specified
e	water vapor pressure	kPa
e_a	vapor pressure at T_{dp}	kPa
e^o	saturated water vapor pressure	kPa
e_s	saturated vapor pressure at $\overline{T}$	kPa
f	Blaney-Criddle factor	mm day^{-1} or mm $month^{-1}$
f(u)	aerodynamic wind function	mm cm^{-2} kPa^{-1} day^{-1}

g	acceleration of gravity	$m\ s^{-2}$
h	enthalpy	$kJ\ kg^{-1}$
k'_c	Blaney-Criddle, consumptive use coefficient	dimensionless
$\hat{k}_c$	Blaney-Criddle, cons. use coef., TR-21	dimensionless
K_t	Blaney-Criddle, climate coef., TR-21	dimensionless
m	month	
p	daylight hours in a year	%
q	specific humidity	dimensionless
r	mixing ratio	dimensionless
r_a	aerodynamic resistance	$s\ m^{-1}$
r_c	crop canopy resistance	$s\ m^{-1}$
s	specific humidity	dimensionless
t	time	hours or days
u_p	wind run at 0.15m above class A pan	$km\ day_{-1}$
u	wind velocity	$km\ day^{-1}$ or $m\ sec^{-1}$
u_*	friction velocity	$m\ sec^{-1}$
x	horizontal distance	m or mm
z	vertical axis	m or mm
z_o	roughness length	m or mm

	Greek Letters	
α	albedo, shortwave reflectance coefficient	dimensionless
α	standard atmosphere lapse rate	$^{\circ}C\ km^{-1}$
δ	sun's declination angle	degrees
Δ	slope of the saturation vapor pressure-temperature curve	$kPa\ ^{\circ}C^{-1}$
ε	emissivity	dimensionless
ε_{bulk}	combined air and earth emissivity	decimal
γ	psychrometer coefficient	$kPa\ ^{\circ}C^{-1}$
γ	specific weight	$kN\ m^{-3}$
κ	von Karman's constant	dimensionless
λ	latent heat of vaporization (enthalpy of vaporization)	$kJ\ kg^{-1}$
θ_z	zenith angle of sun	degrees
ω	hour angle	degrees
ψ	water potential	kPa
ρ	density	$kg\ m^{-3}$
σ	surface tension	$N\ m^{-1}$

GENERAL SUBSCRIPTS AND SUPERSCRIPTS

	superscripts	
$\overline{Y}$	average or mean of Y	
	subscripts	
a	property of air	
a	actual or ambient conditions	
a	power on wind distribution	dimensionless
o	surface, z = zero	
v	vegetation	
z	at elevation z	
2	at an elevation of 2 meters	

AIR TEMPERATURE SUBSCRIPTS AND SUPERSCRIPTS

	superscripts	
$\overline{T}$	average air temperature	K or C
$'$	virtual air temperature	

Subscripts	
a	ambient dry bulb air temperature
(avg)	average or mean air temperature
dp	dew point air temperature
(K)	air temperature in absolute units
(max)	maximum air temperature
(min)	minimum air temperature
wb	wet bulb air temperature
wd	wet bulb depression

AIR HUMIDITY SUBSCRIPTS AND SUPERSCRIPTS

superscripts	
o	saturation conditions

subscripts	
a	property of air
o	surface, z = zero
wb	wet bulb air temperature

EVAPORATION OR EVAPOTRANSPIRATION SUBSCRIPTS AND SUPERSCRIPTS

	superscripts	
$\overline{E}$	average or mean	both E & ET

	subscripts
a	actual or ambient
max	maximum
r	reference
p	potential
(soil)	soil surface

RADIATION SUBSCRIPTS

	subscripts
a	extraterrestrial solar radiation
n	net radiation
nl	net long-wave radiation
s	diffuse and direct solar radiation
so	clear day solar radiation

Chapter 1

INTRODUCTION

The process known as *evapotranspiration*, ET, is of great importance. This interest arises, in part, from academic curiosity about hydrologic processes. Applications of great economic significance include irrigation and water resources studies. An example is the transfer of agricultural irrigation water to municipal and industrial uses. The main reason for applying irrigation water is to supply water for the evaporative process.

Rosenberg et al. (1983) state that 70% of the precipitation striking the earth's surface returns to the atmosphere by evaporation and/or evapotranspiration. They state that in arid regions 90% of the precipitation returns to the atmosphere through evaporative and evapotranspiration processes.

A major article appearing in the August 20, 1990, issue of the US magazine *Time* (Linden, 1990) discusses the global problem of water shortages for all uses. The major use of water is for agriculture according to the article. According to *Time* magazine, wide scale water shortages are so severe in many places in the world that armed conflict over water may erupt in the near future. Major newspapers in the US often feature articles discussing water supplies and consumption.

The discussions that follow involve water use by traditional commercial crops, evaporation from water surfaces, and ET from other plants. Commercial crops are always very important. The importance of evaporation from water surfaces is reinforced.

Most of the physical and biological processes involved in the broadened discussions are the same for natural vegetation and commercial crops. A combined coverage is more efficient with respect to reduced text and ease of understanding.

The target audience for this volume is the practicing engineer or hydrologist. Theoretical treatment is not nearly as complete as found in books written by Brutsaert (1982) or Jones (1992). These books, however, omit many necessary details needed for practical applications. For example, these two books do not discuss reductions in evapotranspiration caused by reduced soil water. The recent ASCE Volume *Evapotranspiration and Irrigation Water Requirements* (Jensen, Burman and Allen eds., 1990) discusses nine versions of combination methods. Only a few versions involving combination theory are covered here for simplicity. Expansion into other variations is left for the reader and other references such as Jensen et al. (1990).

This volume is intended to be complementary to the ASCE reference, (Jensen et al., 1990). While the coverage of combination methods has been reduced here for simplicity, several important additional topics are discussed. These topics include evaporation from water surfaces, crop production functions, and specific flow charts for methods of estimating ET. The ASCE volume includes an extensive comparison between estimates of reference crop ET and lysimeter measurements. Only example calculations are found in this volume.

1.1 Definitions of evapotranspiration and transpiration

The following definitions are written for this volume but are largely compatible with those found in Jensen et al. (1990), ICID (1985), Burman et al. (1980) and Burman et al. (1983). Unless careful definitions are used, it may be impossible to evaluate any work done with the measurement or estimation of ET. See the glossary or list of symbols for more complete definitions.

Evaporation, E, is the process whereby a liquid is transformed entirely into a gas. In this volume water is the liquid, and water vapor is the gas.

Transpiration, T, is the process where liquid water within a plant is transformed into water vapor in the air surrounding the plant.

Evapotranspiration, ET, is the combined process of both evaporation from soil and plant surfaces and transpiration through plant surfaces. The term *consumptive use* is the same as ET with the exception that it includes water needed for the building of plant tissue. The term *consumptive use*, CU, originated in the western US and may be considered to be somewhat archaic.

Maximum evapotranspiration, ET_m, is the maximum ET for a specified crop at a designated time where ET is not limited by dry soil water conditions.

Actual evapotranspiration, ET_a, is the actual ET for a specified crop at a designated time where ET_a has been corrected for possible dry soil water conditions.

Potential evapotranspiration, ET_p, is the rate at which water is removed from the soil surface or profile if available. ET_p has been used in a variety of ways. It always has referred to plants adequately supplied with water and usually not limited by disease or fertility.

Reference crop evapotranspiration, ET_r, is similar to the term ET_p with the exception that it is applied to an identifiable crop, such as alfalfa or grass. ET_r is ET from a well-adapted local crop grown under the same conditions as for ET_p. ET_r is an outgrowth of the confusion surrounding ET_p. It is used as a standard ET for the prediction of crop ET by using crop coefficients.

Crop coefficient, K_c, is a dimensionless ratio used to relate ET_r to ET from a specific crop or plant at a specific time. ET_p has been used with K_c in older publications.

Crop evapotranspiration, ET_a, is ET from a specific plant grown under defined conditions including soil water, fertility, and other cultural conditions. ET_a has been designated as actual or plant ET.

Pan evaporation, E_p, is evaporation from a vessel containing water placed so that evaporation is influenced by the local environment surrounding the pan. E_p has been used as an index of evaporation for reservoirs or ET. Many different kinds of pans have been constructed. Variations include different sizes, materials of construction including color, and whether screened or not. The type of evaporation pan and exposure of the pan must be carefully considered when utilizing E_p data.

1.1.1 *Symbols, variables and units*

S.I. units will be used in most of this volume. Examples of the few exceptions will be some of the equations in Chapter 7. Methods used to predict evaporation from water surfaces are difficult to convert to S.I. units because of the methods complexity and the chance of the introduction of errors. A very few equations will be shown in English units primarily for historic reasons. The description of the Blaney-Criddle method found in Chapter 4 is an example.

Where possible traditional use of variables will be followed with the caveat that definitions common in the US will be used. For example in the US the letter K is used to designate a crop coefficient (see Section 5.4) and in Europe *f* and *g* have been used and called crop factors (Feddes, 1987).

Only the major variables will be shown in the list of symbols. Many minor variables will be defined where they appear in the text. The decision regarding which variables to include is arbitrary and therefore subject to argument.

Many of the parameters calculated in estimating ET have their origin in the science of thermodynamics. Parameters such as specific heat and latent heat of vaporization will have units common to standard thermodynamics texts. For example see the text of Van Wylen and Sonntag (1985).

Reference material specifically addressing the estimation of ET may use latent heat of vaporization in MJ kg^{-1} for convenience while the units kJ kg^{-1} or J kg^{-1} are more common in thermodynamic texts. Standard S.I. thermodynamic units will result when the number 1000 is used to obtain prefixes such as *mega*, M or *kilo* k. When referring to λET the latent heat of vaporization will be converted to MJ kg^{-1} to avoid the repeated multiplication or division by multiples of 1000. See Sections 2.2.11, 8.2.2 and 8.2.3.

To help explain the role of units in estimating ET from climatic data units associated with variables or parameters will be analyzed in the same manner as some recent books dealing with engineering science (Fox and McDonald, 1985). It should be recognized units are subject to the laws of algebra and therefore may be squared or canceled for example. For example, area must be accompanied by the units of m^2 or hectares. Then the division of area in m^2 by depth in m results in a depth in m. Numbers may be changed to desired units by unit analysis recognizing relationships between units. See Sections 8.2.2 and 8.2.3 for more details.

1.2 Measurement of evapotranspiration

The measurement of evapotranspiration from a vegetative surface is a very difficult and time-consuming task. Different approaches to the problem are possible, and most are discussed below. All measurement systems are open to potential criticisms.

In addition to the measurements of ET, climatic data are essential for developing or calibrating empirical methods of estimating ET. Within the last few years, methods of measuring climatic factors have changed from using the *Cotton Belt Instrument Shelter* and manual or strip chart data recording to the use of microprocessor controlled data loggers. These new systems are capable of making many more measurements with greater frequency and are generally capable of partially processing and summarizing data. These systems will store data for periods of a few days to months. The data can then be electronically transferred to a computer for further analysis. This type of data transfer reduces typing or manual transcription errors. Manual transcription of data may reveal faulty sensors or other problems to the operator.

When using an empirical method to estimate ET from climatic data, it is important that the climatic data used in calibrating methods were collected in the same manner as data used in making the operational estimates. For example, data collected within a large moist irrigated area will not be compatible with data collected within a dry desert or that collected at an airport. Unfortunately, airport data are often the only data available.

1.2.1 *Soil water budget methods*

Many of the early measurements of ET were made using the soil water balance methods in fields without lysimeters. At first, gravimetric samples were dried to measure soil water. At the present time, neutron meters are usually used to measure soil water. The method utilizes the following equation for the determination of ET:

$$\mathrm{ET} = P_e + \Delta SM + WT - DR \qquad (1.1)$$

where all units are usually in mm. ET is *evapotranspiration*, P_e is *effective precipitation* (see Chapter 5), ΔSM is the *change in soil water* where a decrease in soil water is entered as a positive number and an increase in soil water is entered as a negative number, WT represents water that may have been *water contributed from a water table*, and DR is *water lost by deep percolation.*

The method provides satisfactory results if measurements are made with sufficient care, and there is reasonable assurance that there is no water transfer from the root zone by deep percolation or water transfer from a water table.

Jensen (1974) recommends the following procedures for satisfactory results.

1. At least 6 sampling locations for fields.

2. The depth to a water table must be much greater than the root zone depth.

3. Only those periods where rainfall is light are used.

Drainage losses are minimized by:

1. Applying the pre-plant irrigation at least 10 days before planting.

2. Applying less water each irrigation than could be retained in the soil.

3. Waiting at least 2 days after an irrigation for soil water measurements if a light irrigation is applied, or longer if a heavy irrigation is applied.

1.2.2 *Soil water measurement*

Soil water measurement is commonly performed but can be very labor intensive and time-consuming. Methods are, on one hand, very simple and, at the other extreme, complex, and new methods are being developed. The topic of soil water measurement is secondary for this volume. Only two common methods are discussed, and the reader is referred to other sources for more information.

Soil water has long been measured by gravimetric procedures. Gravimetric soil water measurement is very simple. Samples of soil are collected and merely weighed in the moist state. After drying in an oven for a specified time, the dry samples are re-weighed and the percent water by weight is calculated. Percent by volume is obtained by multiplying the percent by weight by the apparent specific gravity of the soil under field conditions. Methods of soil analysis (American Society of Agronomy, 1986) is a standard reference showing recommendations. The method has the advantage of simplicity and minimal equipment requirements, but it is labor intensive, and the same location cannot be sampled repeatedly.

Probably the most common method of measuring soil water for uses such as irrigation scheduling and the determination of ET is using the *neutron meter*. The neutron meter utilizes the principle that high energy fast neutrons are slowed by collisions with hydrogen nuclei. Hydrogen nuclei are found in organic material in the soil, but most of the nuclei are found in soil water.

A source of high energy fast neutrons is lowered into an access tube which has been placed in the soil, usually for the life of the study. A detector tube near the source is used to indicate the returning slow neutrons. The method is indirect, and complex electronics are required. In place, field calibration is regarded as being superior, but factory calibration may be utilized. Physical standards are often available. The method is not particularly labor intensive and permits the repeated measurement of the same location within the soil profile. There are dozens of references dealing with access tubing materials and placement and other procedures. Also non-nuclear soil water probes are available for use in access tubes.

1.2.3 *Hydrologic budget methods*

The hydrologic method has often been referred to as the inflow-outflow method. A mass balance equation is used on a watershed, field, or other area.

$$\mathrm{ET} = INFLOW - OUTFLOW + P_e + \Delta SM + WT - DR \quad (1.2)$$

where INFLOW is *measured surface inflow* to a region and OUTFLOW is the corresponding *outflow from an region.* Each term has the units of volume or volume area^{-1}. The remaining variables have been defined in the previous section.

The method has the advantage of determining stream-flow depletion which is the basis for irrigation water rights, but the disadvantage of being unable to isolate ET for a specific crop. It can only give an average ET for the area.

1.2.4 *Lysimetric measurement*

A lysimeter is a tank used to isolate a soil mass containing growing plants (crops) from the surrounding soil. Its purpose is to permit the calculation of a water balance by eliminating percolation beyond the root zone and/or to prevent the addition of water from a water table below the root zone. Lysimeters range from a few centimeters in diameter to as large as 10 meters in diameter. The depth of lysimeters is also quite variable as are the construction materials and methods.

The water balance for a lysimeter is:

$$\mathrm{ET} = P + \Delta SM + water\,added - water\,removed \tag{1.3}$$

where the terms *water added* and *water removed* are self explanatory, and the other terms have been defined in this chapter. Units may be volume or volume per unit area. P is now total precipitation, provided overtopping does not occur.

Lysimeters may be classified in numerous ways but one classification is:

1. Weighing
2. Non-weighing
3. Drainage

Weighing lysimeters usually determine ET by directly measuring a change in weight and converting the weight to ET, assuming water is incompressible. Weighing has been accomplished by commercial scale mechanisms, changes in floating depth for a lysimeter floating in liquid, the use of electronic load cells, changes in pressure of lysimeters resting on pillows containing liquid, and by removing lysimeters from their field locations and then placing them on portable scales.

Non-weighing lysimeters use soil water budget procedures to determine

ET values. The soil water budget procedure has been previously discussed. A specialized soil water budget is required for non-weighing lysimeters.

Drainage lysimeters operate on the assumption that water exceeding ET is applied to the ground surface, and the outflow from a collection point located at the bottom of the lysimeter is measured. ET is then the difference between the water applied and the outflow plus *rainfall.*

At the present time lysimeters, particularly sensitive weighing lysimeters, are regarded as the most reliable method of measuring ET. Lysimeters are not without problems, however. Care must be taken to insure that lysimeters are located well within a sizeable irrigated area surrounded by the same crop as that grown in the lysimeter or at least by a similar crop. An excess area of non-vegetative surface or non-soil surface will alter the heat balance of the lysimeter so that it is not typical of the crop whose ET is being measured. An example of this is an excess area of exposed concrete or steel. According to Jensen et al. (1990) the ratio of the rim area of the lysimeter to the plant area in the lysimeter should be less than 0.1. A lysimeter that is so shallow that normal crop growth is inhibited, or such that the thermal regime within the lysimeter is not typical of the surrounding field, will produce ET measurements that are in error. The proper use of lysimeters is somewhat an *art.*

1.2.5 *Indirect meteorological methods*

Indirect measurements of ET are possible by measuring meteorological parameters and then calculating mass balance or energy balance components. Typical of these methods are the *Bowen ratio* and *eddy-correlation* methods. These methods are becoming more practical because advances in measuring and recording electronic equipment have been dramatic and will continue. Computer facilities for data processing are advancing in technology and are becoming lower in cost. Indirect measurements do not readily provide a season long record of ET. For example, complex vegetation whose ET is being measured may be difficult to identify, especially under conditions of strong advection. The methods do have the advantage of being able to make measurements for some types of vegetation, e.g. phreatophytes not suited for lysimeter measurements.

1.2.6 *Chamber techniques*

ET measurements have been obtained with various types of chambers which are placed over the vegetation. A controlled and known air flow through the chamber is maintained, and the ET rate is determined by measuring the difference in water vapor concentration between incoming and outgoing air. Since any cover over a crop canopy will change environmental conditions, and possibly ET rates, it is usually recommended that measurements be taken soon after the chamber is in place. An example of the use of chambers is (Peterson et al., 1985) where a portable chamber was placed on the front end of a tractor for transport to different locations.

1.2.7 *Biological methods*

There are several methods of measuring water use of plants which provide direct measurements of transpiration rather than ET and are often applied to trees. The methods generally measure the transpiration rate of individual plants, or even individual leaves, rather than of plant canopies. Methods include sap flow techniques, porometers, and potometers. Since the methods measure transpiration rather than ET, they will not be discussed in detail here.

1.2.8 *Passive methods*

Passive methods are considered herein as methods which measure evaporation rates as an indirect measurement of ET rates. The most commonly used passive method is the evaporation pan. Evaporation of water is measured from an open pan, and ET is determined by applying an adjustment factor to the pan evaporation. Other evaporation measurements are obtained with various types of evaporation meters. These usually are some type of wetted paper, cloth, or porous-porcelain device from which evaporation is measured, and ET rates are determined with appropriate adjustments.

1.3 Estimating E or ET using climatic data

Engineers and hydrologists are required to make reliable estimates of both evaporation from water surfaces such as lakes and ponds or evaporation from soil, as well as evapotranspiration involving plants. These plants include both commercial crops such as sugar beets and plants such as willows and trees.

Direct measurements of E and ET are usually too expensive for most undertakings. In addition, the time required for a complete research project often exceeds the time available for the work. The body of knowledge dealing with both E and ET is quite extensive. Verification or checking of results for reasonableness will often permit the use of existing knowledge concerning E and ET.

Flow charts illustrating alternative approaches for the estimation of ET follow (Figures 1.1 and 1.2).

1.3.1 *Estimation of water surface evaporation*

Many of the procedures used to estimate evaporation from water surfaces are unique to the problem itself. However, many of the procedures are specific calibrations of ET methods for the estimation of E. In addition, many of the parameters required are the same for both problems.

For these and other reasons the topic of estimating E appears in a separate chapter. See Chapter 7 which is titled *Evaporation from water surfaces.*

1.3.2 *Direct estimation methods*

Most of the early methods of estimating ET from climatic data involved direct estimation without the intermediate step of a reference crop or the specific use of the word *potential* (see the following for more information). For this reason, methods we have classified as *direct* are discussed in Chapter 5.

1.3.3 *Methods using potential or reference crop ET*

The use of the word *potential* came early in the science of the estimation of ET. *Potential* often referred to soil water limitations but also referred to the type of evaporating surface.

For the purposes of our classification, methods described in this paragraph specifically refer to those methods of estimating ET that utilize the intermediate step of first estimating a potential or reference crop ET. The final step is to estimate the ET for a specific plant at a specific time and place by multiplying the reference crop ET_r by a coefficient K_c which incorporates growth stage and other cultural factors.

The estimation of ET_r is discussed in Chapter 4 and the estimation of ET_a using K_c values is discussed in Chapter 5. Careful explanations of all steps and assumptions are required for the proper evaluation of estimates.

1.3.4 *Methods calculating separate soil evaporation and transpiration*

A unique approach is to estimate *soil evaporation*, E_s, and *transpiration*, T, separately and to then obtain ET estimates by simple addition. This process is discussed in portions of Chapter 5. This approach is used most commonly in hydrologic modeling applications involving the estimation of E_s and ET. The paper by Ritchie (1972) may be considered to be a classic.

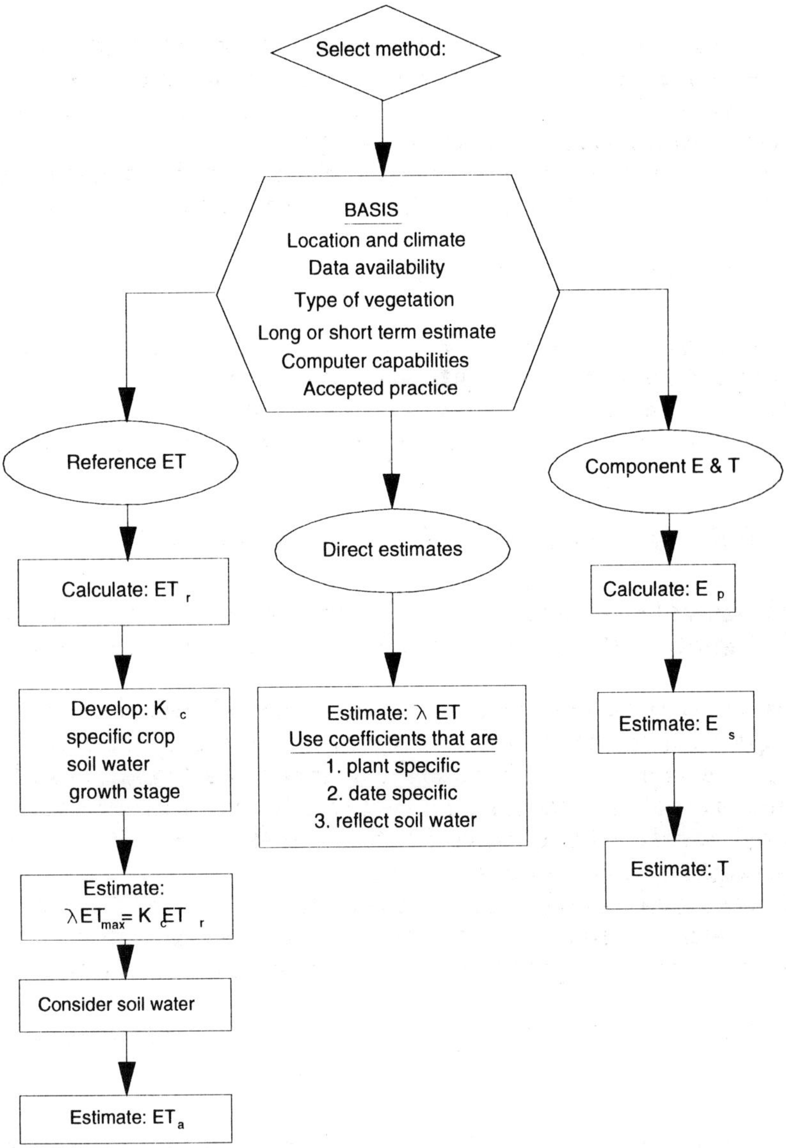

Figure 1.1
General flow chart for estimating ET

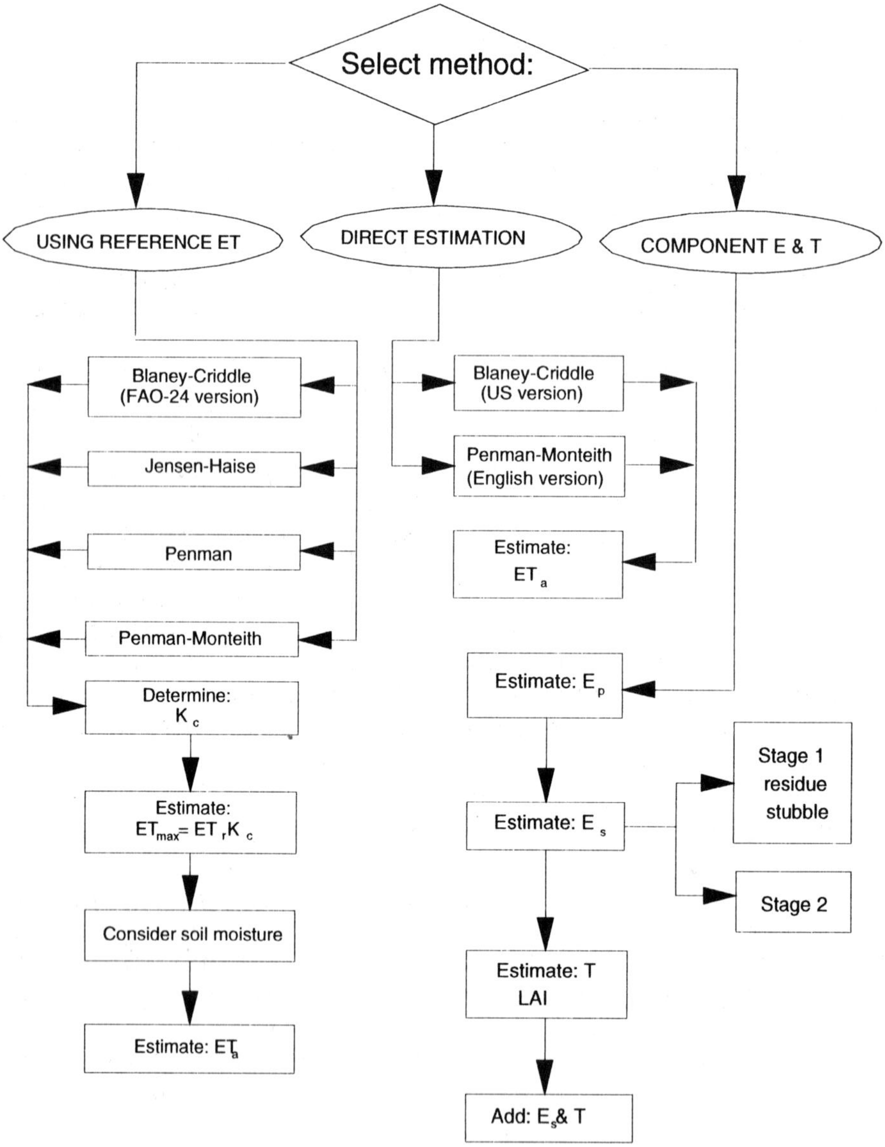

Figure 1.2
Specific flow chart for estimating ET

Chapter 2

ATMOSPHERIC & THERMODYNAMIC PARAMETERS

Many of the methods of estimating ET utilize atmospheric and thermodynamic parameters such as vapor pressure of the air, density of the air, air pressure, and net radiation. These parameters are independent, at least in principle, of the method of estimating ET. However, because of calibration details, some parameters are specific to the method used. The estimation of these parameters is discussed in detail in this chapter, and any modifications related to a particular method of estimating ET are discussed when the method itself is presented.

Methods of estimating ET often utilize specific meteorological parameters for which actual measurements are not available, but which are related to other parameters more commonly measured. For example, net radiation is required for many methods of estimating ET but is seldom measured except in special research related situations. Thus, relationships between net radiation and more commonly measured and/or observed radiation parameters such as solar radiation, cloud cover, and percent sunshine have been developed. Another example is that many methods of estimating ET require some measure of the water conditions of the atmosphere. Thus, relationships between relative humidity, dew point temperature, vapor pressure, etc. are required.

Standard S.I. units will be followed and further, units in common thermodynamics books will be used but modified using S.I. prefixes when needed for convenience. See Section 2.2.11.

2.1 Wind profiles and relationships

Measurement of wind velocity at different elevations above the earth's surface show that the velocity increases with elevation starting at zero velocity at or near zero elevation. The velocity profile near the ground surface is caused by viscous friction. Air is a real fluid and has noticeable viscosity even though less than other common fluids. For example, at 20°C air has a dynamic viscosity of about 1.8×10^{-5} Pa s. Water under the same conditions has a dynamic viscosity of 10^{-3} Pa s, and SAE 30 motor oil has a dynamic viscosity of approximately 0.4 Pa s.

2.1.1 *Time dependency of wind measurements and units*

Everyday experience gained from moving about tells us that wind is usually variable, and at times velocity is dramatically unsteady. In many cases, wind velocity is rather constant when the winds are driven by general circulation patterns. Conversely, both wind direction and magnitude change almost constantly and dramatically with the passage of thunderstorms.

Wind direction variations

On a large scale, wind direction is determined by the locations of air pressure centers and the rotation of the earth. However, factors such as topography can have an influence on wind direction on a local scale.

A basic tenet of meteorology is the concept that air temperature and water content of air is largely determined by surface conditions. Air that originates over water tends to be very moist and is called *maritime*. Air coming from land near the poles is designated as being *polar* and is characterized as being very dry and cold. The extreme conditions cited here are merely to illustrate a point. The temperature and water content of air obviously influence ET, and the direction of wind movement is often of major concern in estimating ET.

McIlroy and Angus (1964) in a classic study reported that ET measured in lysimeters was influenced by whether the wind direction was from the ocean or dry inland areas. Burman et al. (1975) report that wind direction, wind speed, water content, temperature, and estimated ET_p were all changed as air in southern Idaho moved from a rather dry desert environment over a large irrigated area.

While wind direction is of obvious importance in determining ET, no formal means of incorporating wind direction into estimating equations is in use at the current time. In concept, it is possible that different local calibrations might be used depending upon wind direction.

Wind speed variations

Wind speed is almost always measured by the use of anemometers for operational uses, including airport management decisions and the estimation of ET. An anemometer has mass and therefore inertia is important. A response to a wind gust requires time even though it may be small. Anemometer measurements of speed must be considered as non-instantaneous.

Sampling may represent daily intervals in the case of manual recording of wind movement. Many flight facilities record wind movement over hourly intervals. Microprocessor controlled weather stations, now in common use, have sampling intervals as short as one second, but to avoid the storage of unreasonable amounts of data, wind information is usually stored in hourly, quarter day, or daily amounts.

Use and conceptualization of wind data

The units for wind speed may be m s^{-1}, m $hour^{-1}$, m day^{-1} or km day^{-1}. If hourly or daily estimates of ET are being made, data shown in m s^{-1} may be very misleading. The estimator may form the false impression that wind movement was measured over short time periods when it was not. The problem may be amplified when daily averages for a weekly or monthly period are used.

Conversely, when hourly or daily time units are used, the estimator may reach the inaccurate conclusion that wind movement is constant when it is not. An even more serious problem arises because ET occurs mostly in daytime hours, and daily data implies that it is a 24-hour process.

When numerical estimates are made, a careful unit analysis is essential. In the case where the time interval is seconds, a conversion involving the fact that there are 86,400 seconds in a day must be used. The process of unit analysis is discussed in Chapter 8.

2.1.2 *Wind profile relationships*

The most common mathematical description of wind profiles near the ground surface is known as the *log* law. The expression results from an expression for turbulent shear combined with the assumption known as Prandtl's mixing length analogy. In introductory texts dealing with fluid mechanics the resulting equation is

$$u_z = \frac{u_*}{\kappa}\int_{z_0}^{z}\frac{dz}{z} \tag{2.1}$$

where u_* is known as the *friction velocity* because the final units are those of velocity and it is equal to $u_* = \sqrt{\tau_o/\rho}$. The term z_o is known as the *roughness length* in meters and κ is *von Karman's universal constant* which has a commonly used value of 0.41. Table 2.1 shows typical values of z_o for various surfaces. Values of z_0 are broad averages of values from several sources.

The previous expression is valid for pipe and duct flow but does not account for the complexity of wind movement over vegetative surfaces. Agriculturally related micro-meteorologists use the following adaptation.

$$u_z = \frac{u_*}{\kappa}\int_{z_0}^{z-d}\frac{dz}{z} \tag{2.2}$$

where d is defined as the *zero plane displacement* in meters. Figure 2.1 is a sketch showing the various dimensions applied to a field with relatively large vegetation.

The integration of Equation 2.2 results in the well known *log* velocity law.

$$u_z = \frac{u_*}{\kappa} \ln \frac{z-d}{z_o} \tag{2.3}$$

where variables have been defined previously.

Tabular values of d do not appear to be common. Abtew et al. (1989) address the problem of making estimates of both d and z_0. They propose equations for estimating d and z_0.

$$d = 0.85 F_c z_v \tag{2.4}$$

$$z_o = 0.13(z_v - d) \tag{2.5}$$

where d and z_0 have been previously defined and z_v is the average vegetative height in cm. F_c is the fraction of the total surface covered by roughness elements. Presumably d and z_v are also in cm.

Abtew et al. (1989) review 18 empirical equations for the estimation of z_0. The authors know of others. Jacobs and van Boxel (1991) comment that the equations of Abtew et al. (1989) should be used only as a *first guess*.

2.1.3 *Atmospheric stability*

A *neutral* atmosphere for a vertical slice of air is one where temperature, atmospheric pressure, and wind velocity distributions follow adiabatic conditions (no heat exchange). The concept of atmospheric stability is largely based upon the notion of buoyancy. Slightly different criteria are commonly used to determine atmospheric stability. In high altitude meteorology, an observed temperature lapse rate is compared to one calculated using the ideal gas law with the assumption of adiabatic conditions. In detailed micrometeorological studies, the *Richardson Number* is determined by measurements. The *Richardson Number* has been defined by Bartholic et al. (1970) as follows. Other variations may be found; for example, see Brutsaert (1982).

$$Ri = \frac{g}{T}\left[\ln \frac{(z-d)}{z_o}\right](T_a - T_o) u_a^{-2} z \tag{2.6}$$

where Ri is the *Richardson number*, T is air temperature in °C, *g* is the acceleration of gravity in m s^{-2}, the subscript *o* indicates surface conditions, and *a* means in the air. Numerical values of R_i reflect atmospheric stability. When R_i values are positive, atmospheric stability is indicated, and negative values denote atmospheric instability (Van Wijk, 1963).

Table 2.1
Typical values of the roughness length, z_0, for various surfaces

surface	mm avg.	mm low	mm high	remarks
ice, mud flats, smooth	0.01	none	shown	consistent values
water	0.25	0.01	0.6	differs with surface
snow	5	0.05	10	
short grass	10	1	20	
soil surface	11	1	21	
agricultural crops	60	1.5	220	
sugar cane	65	40	90	big crop variations
long grass	70	40	100	big crop variations
corn	100	20	220	big crop variations
orchards	750	500	1000	big crop variations
cities	5000	50	10000	big variations

Figure 2.1 shows that the log law only applies to neutral atmospheric conditions. Logic indicates that the air over vegetation in midday may be influenced by heating from plant or soil surfaces causing buoyancy. The same logic leads to the expectation that air near the ground may be cold and, therefore, relatively heavy in the morning. These conditions would suppress buoyancy. Wind velocity profiles which fit a logarithmic distribution may, therefore, be classified as those of a neutral atmosphere (Figure 2.1).

The majority of *wind functions* used in estimating ET do not make corrections for non-stable atmospheric conditions. Even though formal corrections for stability are not used in estimating equations, any experimentally determined constants appearing in the wind functions would certainly represent the typical conditions occurring under calibration and may, therefore, represent both stable and unstable atmospheric conditions. The proper evaluation of ET estimating methods certainly involves consideration of atmospheric stability conditions.

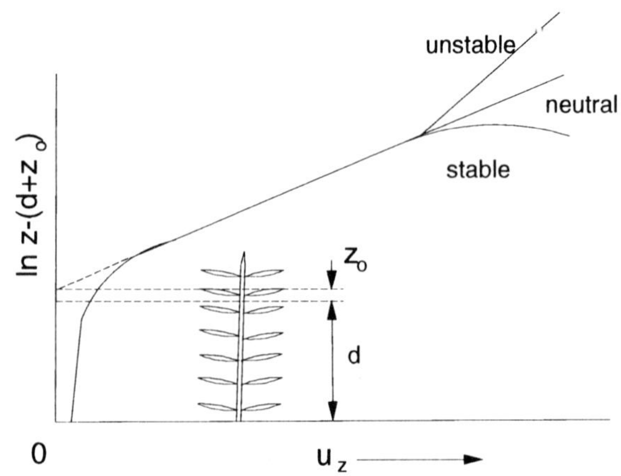

Figure 2.1
Wind speed over plants
idealized logarithmic profile

2.1.4 *Determining wind speed at different elevations*

Wind speed may be measured at different heights than those specified for use in ET estimating methods. A very common occurrence is the case where airport data are the only available wind information. There are procedures for converting wind speeds from one height to another. The standard height for ET calculations is generally 2 meters.

Two expressions are in common use. The first is designated as a power law and has the form:

$$u_z = u_m\left[\frac{z}{z_m}\right]^a \tag{2.7}$$

where z is *elevation above the ground surface* in meters, and the subscript m indicates the level of measurement. The dimensionless exponent a is bounded between zero and unity. Jensen (1974) used a value of 0.2 which was said to represent a rough surface such as alfalfa. When the exponent a has a value of 1/7, the expression is called the Prandtl 1/7 law (Daugherty and Franzini, 1977). This value has often been applied to water flow in pipes. Fox and

McDonald (1985) show a variety of powers based upon the degree of turbulence encountered.

Air flow over vegetative surfaces is much more complex. Rosenberg et al. (1983) show the following expression based upon the log velocity profile law:

$$u_2 = u_1 \frac{\ln\left[\frac{(z_2 - d)}{z_o}\right]}{\ln\left[\frac{(z_1 - d)}{z_o}\right]} \tag{2.8}$$

where all variables have been defined above. The expressions have validity for only a limited distancc abovc thc ground surface where the wind is not influenced by the rotation of the earth.

2.2 Thermodynamic parameters

The thermodynamic parameters used in predicting and explaining ET include, but may not be limited to, atmospheric pressure, air density, specific heat at constant pressure, latent heat of vaporization, and the ratio of the mole weight of water to that of dry air. Other required parameters that may be considered to be thermodynamic parameters include the psychrometer constant and the vapor pressure-temperature relationship. However, the latter two parameters will be discussed in the section covering atmospheric humidity.

2.2.1 *Atmospheric pressure,* P *and density,* ρ

Average atmospheric pressure is needed to calculate the psychrometer constant and air density. For routine estimates of ET, there is no indication that it is necessary to account for the changes in atmospheric pressure which occur with the passage of meteorological fronts. It appears that an average value for a location is sufficient.

Most authors dealing with fluid mechanics, thermodynamics, dynamic meteorology, and hydrology recognize that it is often necessary to be able to estimate an average atmospheric pressure for a given location. These references show tables of atmospheric pressure and density based on the US Standard Atmosphere. The US Standard Atmosphere is intended to represent average conditions at about 40 degrees north in the US. Of interest to workers on the earth's surface is the fact that under average conditions temperature decreases with altitude to an elevation near 11 km where the temperature becomes nearly constant with elevation. The decrease of temperature with elevation is called the *lapse rate*, α. A lapse rate of 6.5 °C temperature decrease per km elevation increase is commonly assumed for the US Standard Atmosphere. The 6.5 °C km^{-1} represents average conditions at mid-latitudes for the US Standard Atmosphere. The lapse rate at the equator is different than that at the poles. The Supplement to the US Standard Atmosphere (Environmental Science Services Administration, 1966) clearly indicates that both the average surface temperature and the lapse rate change with latitude and with season. See Section 8.3.1 for a numerical example.

Equation 7.2 in *Consumptive Use of Water and Irrigation Water Requirements* (Jensen, 1974) was developed by Burman by plotting values of atmospheric pressure against elevation taken from the Smithsonian Meteorological Tables (List, 1984). These tables represent the US Standard Atmosphere (Environmental Sciences Administration, 1966). The plot yielded a straight line relationship with minimal deviations. The slope was determined graphically and the zero elevation intercept was set to a pressure of 101.3 kPa. This expression is easy to use but is difficult to explain without considerable study of the Standard Atmosphere and review of thermodynamics. The user has no method of making modifications for use in a location outside mid-latitudes in the US.

$$P = 101.3 - 0.01055 \cdot z \qquad (2.9)$$

where P is *atmospheric pressure*, kPa, and z is *elevation above sea level*, meters.

The procedure shown in the next section permits the atmospheric pressure to be modified by using different sea level conditions and varying the lapse rate α. The following method can be used to extrapolate data from a location having measurements of atmospheric pressure to another without measured pressure by selecting the appropriate lapse rate and using the measured values.

Constant lapse rate atmosphere

The constant lapse rate atmosphere assumption leads to a very realistic and simple model for predicting the average annual atmospheric pressure for any location using elevation above sea level as the only input. Different lapse rates and base temperatures may be needed for high or low latitudes or other unusual conditions.

The constant lapse rate model results from an integration of a combination of the hydrostatic equation, the constant lapse rate, and the ideal gas law. The ideal gas law is discussed in all thermodynamics texts and represents low density and low temperature gases. Atmospheric air under the pressure and temperature conditions found near the earth fits the ideal gas law very well. The following is a result of this integration:

$$\frac{P}{P_o} = \left[\frac{T_{o(K)} - \alpha(z - z_o)}{T_{o(K)}} \right]^{pwr} \tag{2.10}$$

$$pwr = \frac{g}{\alpha R} \tag{2.11}$$

where α is the *lapse rate of temperature* assumed to be 6.5 °C or K km^{-1} in the US Standard Atmosphere, $T_{o(K)}$ is *air temperature* in K, z is the *elevation of a location* above an arbitrary datum in km, P is estimated *air pressure* at any location in kPa, P_o is the sea level or the pressure at any point where pressure is known again in kPa, R is the *specific gas constant* for day air which has a nominal value of 0.287 kJ kg^{-1} K^{-1}, *g* is the *acceleration of gravity* used as 9.81 m s^{-2}, and *pwr* is a dimensionless exponent defined above. The subscript *o* refers to an arbitrary datum, usually sea level. Equation 2.10 predicts pressure values that are very close to those of the US Standard Atmosphere, or the ICAN Standard Atmosphere. ICAN is an acronym for International Commission for Air Navigation. Even though the specific gas constant for dry air is used, estimates provide reasonable values of pressure for moist atmospheric air.

Both sides of Equation 2.10 are dimensionless. In this volume usually sea level conditions are used. Nominal US Standard Atmosphere sea level values are

$T_{(K)}$	288 K
P_o	101.3 kPa
α	6.5 °C km^{-1}

Air density ρ

Some forms of the *combination* methods for estimating ET require an estimate of air density as an input. Again the ideal gas law may be used. Absolutely dry air is not encountered where estimates of ET are needed. Various methods are available to overcome the problem that the ideal gas law needs modification for use with mixtures of gases. Moist air is treated as a mixture of dry air and water vapor. Moist air is actually lighter than dry air because of light hydrogen atoms. The concept of *virtual temperature* is used. The virtual temperature is the dry bulb temperature corrected for water so that the specific gas constant for dry air will yield the correct density for moist air. Equations used in the virtual temperature correction are.

$$\rho_m = \frac{P}{RT_{(K)}'} \tag{2.12}$$

$$T_{(K)}' = \frac{T_{(K)}}{1 - 0.378\frac{e_a}{P}} \tag{2.13}$$

where ρ_m is the *density of moist air* in kg m^{-3}, T is *air temperature* in K, *e* is the *partial pressure* resulting from water vapor in the air in kPa, P is *atmospheric pressure* in kPa, and R is *the specific gas constant* for dry air. T′ is known as the *virtual temperature* and is the air temperature adjusted so that the ideal gas law can be used to predict the density of moist air using the specific gas constant for dry air instead of a weighted specific gas constant for both dry air and water vapor. A common value of the specific gas constant, R, for dry air is 0.287 kJ kg^{-1} K^{-1}. The term *e* / *P* becomes an approximation of the specific humidity when multiplied by 0.622 which is the ratio of the mole weight of water vapor compared to the mole weight of air.

2.2.2 *Atmospheric humidity calculations*

It is obvious that dry air has more of a tendency than moist air to evaporate water from soil and water surfaces and to transpire water. Some methods of estimating ET, such as those using only air temperature, do not include any terms involving the water content of air. To a major extent, this omission is due to a lack of such data. More complete methods, such as *combination methods*, do require a knowledge of the water content of the air. The widespread use of relatively low cost microprocessor controlled climatic stations is now, and will continue to be, a major reason for including the water conditions of the evaporating air in ET estimations.

The water content of the air can be expressed using a number of different parameters including vapor pressure, relative humidity, dew point temperature, mixing ratio, specific humidity, and absolute humidity. Of these, the first three are normally used in ET formulas. Often, however, the parameter measured by a weather station is not the same parameter as that required by a specific formula. Thus, methods for converting from one parameter to another must be available.

2.2.3 *Vapor pressure* e

All liquids, including water, tend to vaporize into air as a result of molecules leaving the surface of the liquid. The molecules of the liquid which has become a gas exert a partial pressure, *e*, known as *vapor pressure*. At some time, equilibrium is reached between the molecules escaping and reentering the liquid. At this time the partial pressure is known as the *saturation vapor pressure*, e^o. The saturation vapor pressure e^o increases in a non-linear rate with increasing air temperature.

Saturation vapor pressure values shown in the Smithsonian Meteorological Tables (List, 1984) are often considered as standard (Brutsaert, 1982). An even better standard may be data appearing in thermodynamic properties of water known as *steam tables* (Keenan et al., 1979). These data appear in all thermodynamics texts and have been revised and updated many times. Steam table data are regarded as a standard by many industries.

Many empirical formulae relating saturation vapor pressure to air temperature have been developed. Depending on the purpose intended, the formulae are of varying complexity. The meteorologist works with temperatures generally below 50 °C while thermodynamic tables of properties for water vapor have saturation vapor pressure data for temperatures to about 750 °C. Empirical models used by meteorologists, therefore, can be simpler because they only have to duplicate values found in steam tables over a narrow range.

Much of the interest in developing alternate models for estimating vapor pressure in the atmosphere was generated by a need to have models that could be used repeatedly in calculations with a minimum of computer time as a part of large atmospheric models used for weather prediction. Burman et al. (1987) and Weiss (1977) show that any one of several methods of estimating vapor pressure can be used for the estimation of ET from the standpoint of accuracy. An additional reason that many satisfactory models are available is the fact that computer time is not important in estimating ET because repetitive calculations are usually not a problem.

The authors show in Chapter 8 that a form of an equation by Tetens' (1930) equation credited to Murray (1967) is satisfactory for estimating vapor pressure. Weiss (1977) discusses this equation.

$$\frac{e^o}{e^o_{so}} = \exp\left[\frac{17.27T}{T + 237.3}\right] \qquad (2.14)$$

where e is in kPa, the superscript o indicates saturation vapor pressure, the subscript so indicates saturation vapor pressure at 0°C, and T is in °C. For the units used in this volume $e^o_{so} \sim 0.6108 kPa$. The equation is very satisfactory up to an air temperature of about 100 °C.

2.2.4 *Slope of vapor pressure and temperature relationship* Δ

For all versions of the combination (Penman) equations and some of the methods of estimating E using a form of solar energy, numerical values of Δ need to be calculated. Any satisfactory relationship used to estimate vapor pressure may be differentiated with respect to temperature to obtain an algorithm for the estimation of Δ. Differentiation of the form of the Tetens' equation, shown in the previous section, results in the following:

$$\Delta = \frac{2503}{(T+237.3)^2}\exp\left[\frac{17.27T}{(T+237.3)}\right] \tag{2.15}$$

where Δ is in kPa °C^{-1}.

2.2.5 *Relative humidity RH*

The term *relative humidity* is commonly used with respect to climatic considerations, but its definition is not as simple as it appears. For a constant air temperature and pressure, *relative humidity*, RH, is defined as the ratio of the actual mixing ratio to the saturation mixing ratio (Hess, 1959). Assuming water vapor is an ideal gas, this definition leads to the following (Van Wylen and Sonntag, 1985):

$$RH = e_a / e^o_{T_a} \tag{2.16}$$

where e is *vapor pressure* in kPa. The superscript o represents saturated conditions, e_a is the *vapor pressure of the air* in kPa, and the subscript T_a represents air temperature at ambient conditions. RH may be expressed as a dimensionless ratio or as a percentage.

RH is temperature dependent and may be very misleading as an indicator of the water content of air. For example, the water content of saturated air at cold conditions such as freezing or below is small while the water content of low RH air at high temperatures such as 40°C will usually be much higher.

2.2.6 *Mixing ratio r*

The mixing ratio is defined as the ratio of the mass of water vapor in a parcel of air to the mass of dry air.

$$r = \frac{\rho_v}{\rho_d} = 0.622\frac{e_a}{P - e_a} \tag{2.17}$$

where r is the *mixing ratio* and is dimensionless, P is *atmospheric pressure* in kPa, e_a is the actual *vapor pressure* of air in kPa, ρ_v is the *density of vapor in the air* in kg m^{-3}, and ρ_d is the *density of dry air* in kg m^{-3}.

2.2.7 *Specific humidity q*

Specific humidity, q, is the ratio of the mass of water vapor in a unit volume of air to the mass of moist air.

$$q = \frac{\rho_v}{\rho} = 0.622 \frac{e_a}{P - 0.378\, e_a} \tag{2.18}$$

where ρ_v is the *density of water vapor in the air* in kg m^{-3}, ρ is the *density of moist air* in kg m^{-3}, e_a is the *vapor pressure of the air* in kPa, and P is the *total atmospheric pressure* in kPa. For most cases, the mixing ratio is nearly equal to the specific humidity.

2.2.8 *Psychrometer measurements and equation*

One of the earliest methods of measuring the water content of the air was the sling psychrometer. In spite of the fact that many methods are available for measuring water content in the air, the sling psychrometer is still used as an independent check on more sophisticated methods of measurements. The simple sling psychrometer is merely a pair of thermometers mounted on a bracket which is attached to a chain. The psychrometer may be rotated through the air by swinging the chain holding the psychrometer.

A pair of thermometers is mounted on the psychrometer. One of the thermometers has a muslin cover which is saturated. The other is not covered, and its temperature measurement is known as the *dry bulb temperature*, shown here as T_a. As the psychrometer is moved through the air, the temperature of the moist bulb is cooled by evaporation. The psychrometer is swung through the air until the temperature on the thermometer with the wet muslin remains constant, which is known as the *wet bulb air temperature*, T_{wb}.

Knowing T_a and T_{wb} plus the elevation of the point above sea level, it is easy to determine the vapor pressure and all other important parameters which describe the water content of the air. The classic equation known as the *psychrometric equation* is used as a basis for the calculations and provides a means of explanation for many of the air-water parameters which follow:

$$\Delta T_{wd}(\rho\, C_p + \rho_v C_{pv}) = (\rho_s - \rho_v)\, \lambda \tag{2.19}$$

$$\Delta T_{wd} = T_a - T_{wb} \tag{2.20}$$

where ΔT_{wd} is the *wet bulb depression* $T_a - T_{wb}$ in °C or K, T_a is the *dry bulb air temperature* °C or K, T_{wb} is the *wet bulb air temperature* °C or K. ρ is the vapor density of the air. The subscripts *v* means entering air and *s* indicates saturated air leaving the bulb. λ is the *latent heat of vaporization* kJ kg^{-1}. This equation is from Eagleson (1970).

The psychrometer equation is often simplified as:

$$\Delta T_{wd}\gamma = e^o_{wb} - e_a \tag{2.21}$$

where γ is defined here as the *psychrometric coefficient* kPa $^\circ C^{-1}$, e^o_{wb} is the *saturation vapor pressure* at the wet bulb air temperature in kPa, and e_a is the actual *vapor pressure* of the ambient air. This simplification assumes that a suitable C_p for moist air be used. Previously, the term γ was defined as the *psychrometric constant.* The use of the word *constant* is misleading because γ is indeed a variable. The psychrometric coefficient, γ, is discussed in detail in *Psychrometric coefficient* γ. There may be variations in the *psychrometer equation*, depending on units used and simplifying assumptions. See Chapter 8 for one derivation of the *psychrometric coefficient.*

2.2.9 *Actual vapor pressure* e_a

The actual vapor pressure of air may be determined by the use of a sling psychrometer and the use of the psychrometer equation discussed in the previous section. In the case of a hygrothermograph, or other instruments such as microprocessor controlled automated weather stations, the following expression may be used:

$$e_a = e^o_{dp} = RH \cdot e^o_{db} \tag{2.22}$$

where e_a is the actual *vapor pressure* of the air in kPa, and e^o_{dp} is the *saturation vapor pressure* in kPa at the dew point temperature in °C. Values of *e* are in kPa. The subscript *dp* indicates the temperature is at the dew point, which is defined as the temperature where the ambient air is saturated when it is cooled at constant moisture content. The subscript db indicates the dry-bulb air temperature. RH is the *relative humidity* and is used either as a decimal or percentage depending on the specific equation used.

2.2.10 *Specific heat at constant pressure* C_p

Specific heat, C_p, at constant pressure is defined as the amount of heat required to change the temperature of a unit mass one degree at constant pressure, kJ kg^{-1} K^{-1}. Specific heat at constant pressure, C_p, for dry air is commonly used with a value of 1.004 kJ kg^{-1} K^{-1} at standard sea level conditions. This value varies slightly in the fourth significant digit, depending on the reference used. Brutsaert (1982) recommends the following method of estimating a weighted specific heat for moist ambient air.

$$C_{pm} = (1 + 0.84q)C_{pda} \tag{2.23}$$

where the subscripts on C_p of *pm* indicate moist ambient air, and *pda* means completely dry air, and *q* is specific humidity which is a dimensionless ratio. Fleagle and Businger (1980) show the same equation but with the constant having the value of 0.9 instead of 0.84.

The use of an average value of C_p for moist air is justified because C_p varies only slightly, and ET estimates are not sensitive to changes in C_p. To be correct theoretically, a value of C_p for moist air should be used. The calculation of C_p for moist air is not difficult using computer algorithms. The authors recommend an average value of C_p of 1.01 kJ kg^{-1} K^{-1} be used for ET determinations based on a table appearing in Section 8.7.6

2.2.11 *Latent heat of vaporization λ*

Latent heat of vaporization, λ, is the heat required to change a unit mass of water from a liquid to a water vapor in a constant pressure and constant temperature process and commonly has the units of kJ kg^{-1}. Latent heat of vaporization is used in calculations of the *psychrometric coefficient*, γ, and in converting energy values of ET to volumes of water. Errors in estimating λ do not appear to be important in calculations of γ but could be of significance in converting energy values of ET to equivalent volumes of water.

In thermodynamic references, λ is known as enthalpy of vaporization (Silver and Nydahl, 1977; Van Wylen and Sonntag, 1985). Values of λ are found in steam tables under the columns for enthalpy. Specifically, λ is h_{fg} which is the change in enthalpy required to move from the saturated liquid point to the saturated vapor point within the steam tables.

Latent heat of vaporization may be estimated using the following linear regression equation based on thermodynamic steam tables (Harrison, 1963):

$$\lambda = 2501 - 2.3601 \cdot T \tag{2.24}$$

where λ has been described above, and T is *air temperature* in °C. Burman et al. (1987) recommend that T be used as the average daily air temperature in normal ET calculations, and that T should be the wet bulb air temperature in making psychrometric calculations.

In chapters 2,4 and 8 the units for λ of kJ kg^{-1} will often be converted to MJ kg^{-1} to simplify calculations.

2.2.12 *Psychrometric coefficient γ*

Most references define *gamma*, γ, as the psychrometric constant. In this case γ is defined as the *psychrometric coefficient* to reinforce the conclusion that γ is not a constant in reality. Gamma is, in actuality, strongly dependent on elevation and weakly dependent on temperature.

$$\gamma = C_{pm} P \,/\, \varepsilon\lambda \tag{2.25}$$

where C_{pm} is *specific heat at constant pressure* for naturally occurring moist air in kJ kg^{-1} K^{-1}. Equation 2.23 can be used to estimate a weighted value of C_{pm} for moist conditions. The authors show that C_p has a value of about 1.010 kJ kg^{-1} K^{-1} for common conditions encountered when estimating ET. See Section 8.7.6 for a discussion of specific heat at constant pressure and selected common values of specific heat at constant pressure. P is *total atmospheric pressure* in kPa, and λ is the *latent heat of vaporization* in kJ kg^{-1}. λ is discussed in the previous section. The constant ε is the ratio of the mole weight of water to air and has a value of 0.622.

2.3 Radiation

Heat is transferred by conduction, convection, and radiation. Radiant heat transfer is very important in ET because the source of energy for ET comes from the sun. A knowledge of radiant heat transfer and the ability to estimate solar and net radiation is vital for the prediction of ET.

Radiation transfers energy by means of electromagnetic waves characterized by wavelength or wave-number. As far as water use calculations are concerned, the wavelengths of importance are those in which the sun, the earth, and the atmosphere radiate. Wein's Displacement Law shows that the wavelength of maximum intensity is an inverse function of temperature (Welty et al., 1976). For this reason, radiation from the sun, which is at a much higher temperature than the earth, has a much shorter wavelength or maximum intensity than radiation from the earth.

Solar radiation occurs in the so-called *short* wavelengths of about 0.15 to 4 microns with a maximum intensity near 0.5 microns. Radiation from the earth and atmosphere, that is terrestrial radiation, is referred to as *long wave* radiation and is mainly in the 4 to 50 micron wavelengths with a maximum intensity near 10 microns (Gates, 1962). There is very little overlap in wavelengths between solar and terrestrial radiation. Solar radiation reaching the earth's surface is limited essentially to wavelengths above 0.3 microns because the shorter ultraviolet wavelengths are screened out by the ozone layer of the atmosphere. Solar radiation includes the visible spectrum of approximately 0.4 to 0.74 microns.

The rate of radiant heat emission is given by the Stefan-Boltzmann equation, or simply Stefan's equation.

$$\frac{q}{A} = \varepsilon' \sigma T^4 \tag{2.26}$$

where q is *heat transfer* in energy units MJ day^{-1}, *emissivity*, ε' is the *ratio of the radiation from the surface considered to that from a "perfect" radiation surface* where ε' has a value of unity, A is the *area* emitting radiation m^2, T is *absolute temperature* in K, and σ is known as the *Stefan-Boltzmann constant* which in this case has a value of $4.90 \cdot 10^{-9}$ MJ day^{-1} m^{-2} K^{-4}.

2.3.1 *Solar radiation* R_s

Many estimates of ET involve historic climatic records which usually do not include solar radiation data. Therefore, it is often necessary to estimate solar radiation from other data. Solar radiation may be measured on a real time basis with comparative ease using microprocessor controlled data acquisition systems or estimated from related measured data using an appropriate model. Discussion here is limited to daily or monthly estimates of solar radiation and to those methods which seem to be useful on a practical level.

2.3.2 *Vertical energy balance*

Energy from the sun in the form of solar radiation striking the earth's surface is accountable in an energy budget. Figure 2.2 shows the disposition of radiant energy at noon on a summer day for a soil-plant surface.

2.3.3 *Solar constant SC*

Solar radiation at the top of the earth's atmosphere on a surface perpendicular to the sun's rays varies with time but is constant enough for many applications to justify use of the term *solar constant*. This term is quite useful in estimating both *extraterrestrial solar radiation*, R_a, and *clear day solar radiation*, R_{so}. The value of the *solar constant*, SC, has been estimated on the basis of high altitude measurements to be from 1353 to 1396 W m^{-2}. A recent value is 1367 W m^{-2} (Frolich, 1982). The value of 1367 W m^{-2} is equal to 118.1 MJ m^{-2} day^{-1}. Minor variations are not significant from a practical standpoint in estimating ET.

2.3.4 *Albedo* α

Albedo, α, is defined as *reflecting power* in a common dictionary (Thorndike and Barnhart, 1974) and more specifically as *the ratio of electromagnetic radiation reflected from a soil and crop (plant) surface to that incident upon it* as defined by ASCE (Jensen et al., 1990). Primarily albedo refers to a somewhat narrow definition of a portion of the range of *fractional reflectance of radiation*. Albedo has been described as referring to visible or solar radiation. In this document, albedo relates to the total reflectance of solar radiation both diffuse and direct (Pielke, 1984). Albedo varies with the kind of plant and, in some cases, notably for water or snow, is highly dependent upon the angle of incidence. Albedo is used in calculating energy balance quantities and, in particular, albedo is used in models to predict *net radiation*.

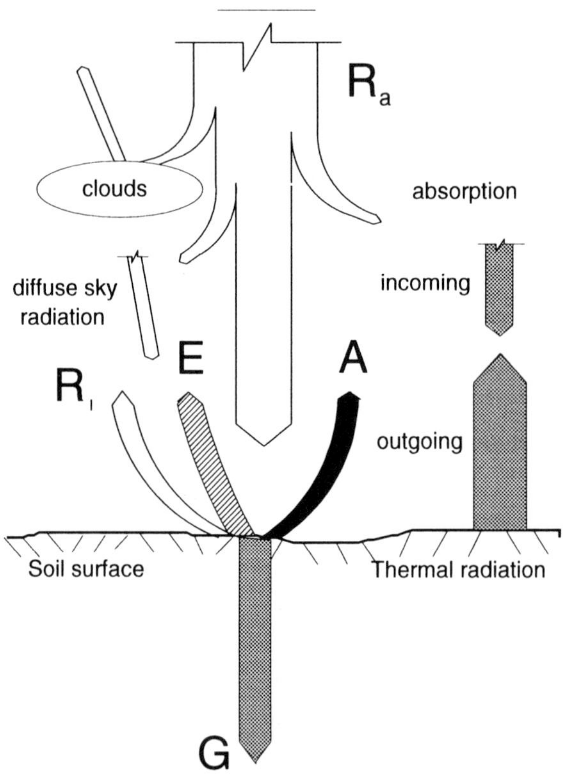

Figure 2.2
Sketch showing the heat balance at solar noon.

In the sketch R_a indicates extraterrestrial solar radiation, R_l means long-wave radiation form the earth, E means evaporation, A designates heating of the air and G means heat transfer to the soil. All have the units of MJ m^{-2} day^{-1}.

2.3.5 *Constant albedo*

From common experiences, it is easy to predict that the albedo of a plant will vary throughout its growing season. In most cases, knowledge limitations force the estimator to use a constant α for plants throughout the growing season. Table 2.2 is a summary of α values for plants and other surfaces. These values of α represent broad averages from several sources. Jensen et al. (1990) recommend a value of $\alpha = 0.23$ for alfalfa used as a reference crop, and Doorenbos and Pruitt (1977) recommend a value of $\alpha = 0.25$ for a grass-reference crop.

Table 2.2
Typical values of albedo α
from a variety of sources

surface	genl. avg.	low	high	remarks
water, general	0.50	0.02	1.00	depends on sun's angle
clouds	0.32	0.05	0.84	kind & configuration
snow	0.63	0.29	0.95	depends on age
urban & cities	0.16	0.12	0.21	
streets & roads	0.19	0.10	0.28	
roofing	0.13	0.12	0.14	
soils	0.18	0.05	0.31	
forest coniferous	0.18	0.10	0.24	
forest deciduous	0.18	0.15	0.20	
alfalfa	0.23	0.20	0.32	varies thru growing season
grass	0.25	0.14	0.45	ditto
corn	0.18	0.16	0.23	ditto
small grain	0.19	0.10	0.26	ditto
cotton	0.21	0.20	0.22	ditto
fallow	0.08	0.05	0.12	ditto
potatoes	0.19			ditto
sugar beets	0.18			ditto

2.3.6 *Variable albedo model*

Recognizing that α varies throughout the season, Wright (1982) developed the following for the reference crop alfalfa:

$$\alpha = 0.29 + 0.06 \sin[30(m + 0.0333\, N + 2.25)] \tag{2.27}$$

where N is the *day number* within the month and m is the *month number*. When using Wright's variable albedo model, the Wright variable wind function should also be used and Wright's *crop coefficients*, K_c as well. Although Wright's model was developed for conditions found at Kimberly, Idaho, for alfalfa used as a reference crop, the model has apparently been used on a widespread basis.

2.3.7 *Sun-earth geometric relationships*

General relationships

The need for expressions to estimate various components of both shortwave radiation and terrestrial radiation has been stated. Because the ultimate source of energy for ET is the sun, many of the calculations utilize trigonometric functions. These trigonometric functions are based on geometric relationships involving the earth and sun.

Most of the required trigonometric relationships appear in a wide variety of sources. These include cadastral surveying, climatological, meteorological and hydrologic references, particularly book length volumes. Typical of the diverse references is the discussion of the earth-sun relationships presented in a classic surveying book by Davis and Foote (1953). Most of the equations, therefore, do not appear to be creditable to individuals or groups but rather appear to be a part of the general body of knowledge. The exceptions appear to be empirical functions used to estimate specific quantities, in which case specific individuals will be recognized. There appears to be a wide variety in the symbols used to define angles and quantities. In general, the nomenclature of Garg (1982) will be followed.

Figure 2.3 is a definition sketch showing that the earth is tilted with respect to the sun. The distance between the earth and the sun changes slightly throughout the year. These facts must be an integral part of any equations used for the prediction of radiation components.

Two geometric expressions that are used in models useful in predicting ET are the relationships that define the *zenith angle*, θ_z of the sun, and the relationship for *the hour angle*, ω_h. Garg (1982) and Sellers (1965) discuss these terms. The zenith angle θ_z is defined as

$$\cos\theta_z = \sin L \sin\delta + \cos L \cos\delta \cos\omega_h \tag{2.28}$$

A general expression for the hour angle ω_h is

$$\cos\omega_h = \sin\left[\frac{\cos\theta_z}{\cos\delta\cos L}\right] - \tan\delta\tan L \tag{2.29}$$

where θ_z is the *zenith angle of the sun*, δ is the *declination of the sun*, and L is the *latitude* in degrees. Because the $\cos\theta_z$ is zero at sunset or sunrise, the hour angle at sunrise and sunset reduces to the following:

$$\cos\omega_{hs} = -\tan\delta\tan L \tag{2.30}$$

where ω_{hs} is known as the *sunset hour angle*. The hour angle is then the angle between the sun at solar noon and the sun at sunrise or sunset and is in degrees.

Time relationships

The forgoing relationships apply to solar time. However, the world has been arbitrarily divided into time zones where time is based on standard clock time over a latitude range. The United States time zones are sized so that they differ by even hours. Some countries have time zones that differ by 1/2 hour intervals.

People considering evaporation need to be able to convert local time to solar time. Literature dealing with time relationships is found in many seemingly different areas such as the solar heating and surveying disciplines.

Based upon purely geometric considerations 1 hour is equal to 15°, 360° of longitude is equal to 24 hours. The general time relationship follows:

$$local\ app.\ time = local\ std.\ time + longitude\ corr. + eq.\ of\ time \tag{2.31}$$

where the *local apparent time* may be considered to be solar time, the *local standard time* is the local clock time based upon time zones and ignoring any daylight savings time and the longitude correction is needed because nearly all points on the world are not at an exact time zone change and the *equation of time*, EOT, is needed because the actual day length is not an even 24 hours in length. Times in the above equation are usually in minutes.

In minutes the longitude correction is

$$longitude\ corr = 4(L_e - L_s) \tag{2.32}$$

where L_s is the time zone or standard longitude and L_e is the local longitude of the point. The longitude correction is in minutes of time.

The mean day is 24 hours of time but the actual day varies throughout the year because the earth's orbit about the sun is elliptical and not a perfect circle, the earth's axis is tilted with respect to the sun, and the earth tends to wobble about its axis. This variation has a maximum value of about 14 minutes. For precise work such as surveying, values of EOT should be obtained from ephemeris tables for a specific year. For other work the EOT may be estimated from an empirical equation. The following periodic equation is credited to Woolf (1968). Other such equations are available.

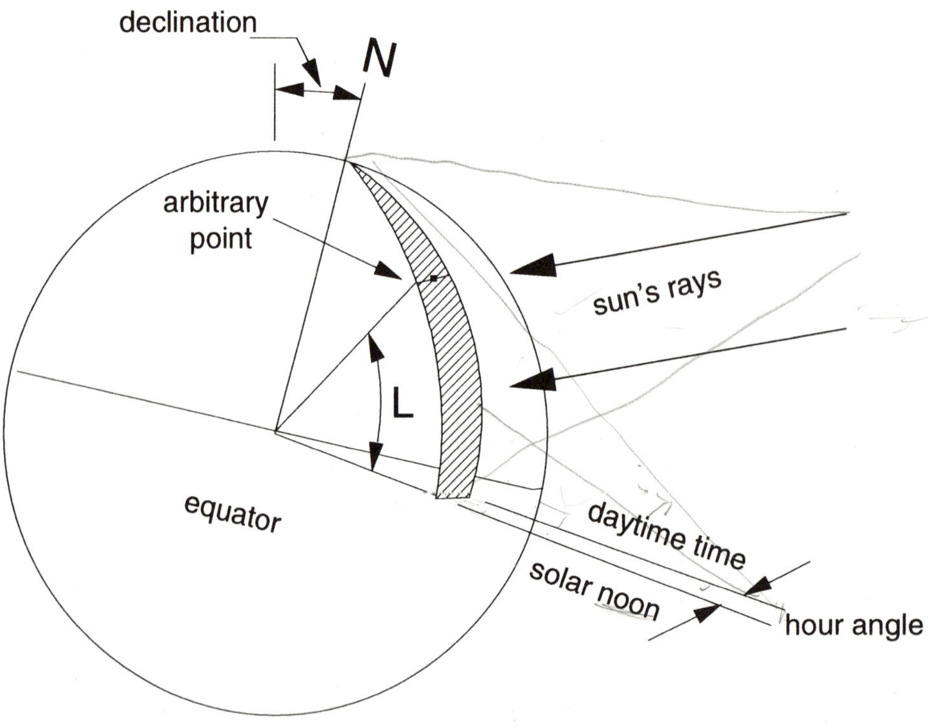

Figure 2.3
Sketch showing earth-sun geometry, northern hemisphere, summer conditions

$$EOT = 0.258 \cos\Gamma - 7.416 \sin\Gamma - 3.648 \cos 2\,\Gamma - 9.228 \sin 2\,\Gamma \quad (2.33)$$

where

$$\Gamma = \frac{360}{365.242}(D - 1)$$

Table 2.3
US standard time zones and longitudes

time zone	meridian
Atlantic	60
Eastern	75
Central	90
Mountain	105
Pacific	120
Alaska	135
Hawaii	150

all meridians are west
all times are standard
meridians are in degrees

2.3.8 *Declination of the sun* δ

The earth rotates about an axis defined by the north and south poles. The polar axis of the earth, however, is inclined with respect to the sun at a maximum angle of about ± 23.5 degrees. In reality, the earth orbits about the sun, but in many analyses it is more convenient to view geometry as if the earth is stationary, and the sun moves with respect to the earth. This agrees with our observations on the surface of the earth. Many analyses of fluid mechanics assume the observer is moving with the fluid. Because the axis of the earth is tilted with respect to the sun, the familiar seasons of year are defined by the apparent angle of the earth with respect to the sun. In the summer solstice the angle of the earth with respect to the sun, *solar declination*, δ, is at a positive angle of about 23.5 degrees. Conversely, the winter solstice has a δ of a negative 23.5 degrees. The declination, δ, is zero at two times in the passage of a year: the autumnal equinox and the vernal equinox.

Precise information dealing with δ may be found for individual years in volumes known as a *solar ephemeris*. The precise values of δ are required for purposes such as the determination of the true bearing of a line determined from sights on the sun and for other purposes such as navigation.

The *declination of the sun*, δ for a given day of the year varies from year to year because of the fact that the time required for the earth to make one revolution around the sun is about one-fourth of a day longer than a year. This leads to the necessity of including an extra day in February every four years.

Numerical values of δ are required to estimate day length and radiation parameters such as extraterrestrial radiation. While precise values of δ are required for procedures such as surveying, the small yearly variations are not of practical significance in applications such as solar collector design or the estimation of evaporation.

At least three algorithms exist for estimating δ for any desired calendar day of the year. Results of making estimates of δ using the three methods were compared with values of δ shown in a joint publication of the US Navy and Royal Navy of Great Britain for the year 1990 (United States Naval Observatory, 1988b). The results of these comparisons are shown in Chapter 8. In addition, equations for estimating δ were developed from the 1990 data and are proposed for operational use because of a combination of simplicity and improved accuracy in estimating δ for 1990. However, it is assumed that the model of Burman and Jacquot, described below, is satisfactory for routine use in estimating ET because yearly variations of δ are of small.

2.3.9 *Cooper declination model*

The *declination of the sun*, δ, may be estimated as follows (Cooper, 1969).

$$\delta = 23.5 \sin\left[\frac{360(D+284)}{365}\right] \tag{2.34}$$

where δ is in degrees, and D is the *calendar day number* for the date in question. The expression of Cooper is in the most widespread use of any method and is quite accurate during the growing season but is in error of over a degree in October. See Chapter 8 for evaluations of estimate accuracy.

2.3.10 *Burman-Jacquot declination model*

A procedure developed by Jacquot and Porter (1989) for the Fourier analysis of periodic functions was used to develop a cosine series to predict δ. Three terms of the series were evaluated. The use of the two terms shown below are recommended because the algorithm is only slightly more complicated than that of Cooper, and the two terms will more accurately represent the δ values shown in the 1990 ephemeris and presumably δ values for other years. The first two terms of the series follow:

$$\delta = 0.3931 + 23.2577 \cos(0.9861\,D - 170.7) + 0.3906 \cos(1.9154\,D - 174.4) \tag{2.35}$$

where D is the *calendar day number* of the year.

2.3.11 *Variations in the sun-earth distance* $\left[\frac{\bar{d}}{d}\right]^2$

The distance between the sun and the earth is known to vary slightly during the passage of a year. The earth-sun relative distance is also shown in astronomical tables for individual years. For many common engineering purposes such as the design of solar collectors or estimating evaporation, variations between years is not significant. For this reason simple empirical algorithms have been developed.

2.3.12 *Duffie-Beckman sun-earth distance correction*

A simple periodic function credited to Duffie and Beckman (1980) follows. Corrections are usually of the form $\left[\frac{\bar{d}}{d}\right]^2$ because the intensity of radiation is known to be inversely related to the square of the distance between radiating bodies.

$$\left[\frac{\bar{d}}{d}\right]^2 = 1 + 0.034 \cos\left[\frac{360\,D}{365.25}\right] \tag{2.36}$$

where D is the *calendar day number* of the day, $\bar{d}$ is the average distance from the earth to the sun, and d is the actual distance between the earth and the sun.

A more complex model intended to predict the earth-sun distance correction was developed by Spencer (1971) using Fourier analysis techniques. The Spencer and Duffie-Beckman models are evaluated in Chapter 8. Results of comparisons shown in Chapter 8 indicate that the Spencer model is slightly more accurate in its predictions. However, based upon performance and convenience, the Duffie-Beckman model is recommended.

2.3.13 *Day length predictions*

Numerical values of the day length in ratios of 12 hours were required to use the Thornthwaite (1948) method of estimating evapotranspiration. The Thornthwaite method is not used at the current time except as a climatological parameter and, therefore, is not discussed in this volume. Day length, however, is used in the determination of the percent daytime required for the Blaney-Criddle method which is discussed in Chapters 4 and 5 of this volume. The number of daylight hours in any day is simply

$$H = \left[\frac{2 \cdot 24}{360}\right]\omega_{hs} \tag{2.37}$$

where H is in hours, and the sunset hour angle, ω_{hs}, is in degrees.

2.3.14 *Percent daytime hours* p

Monthly or daily values of the percentage of daytime hours on an annual basis are required for the various versions of the Blaney-Criddle method of estimating evapotranspiration. The monthly values were calculated as follows:

$$H_i = H_{mid} DAYS_i \tag{2.38}$$

$$p_i = 100 \cdot \frac{H_i}{\sum_{i=1}^{12} H_i} \tag{2.39}$$

where p is a percentage, the subscript *mid* indicates the day nearest the mid-point of a month, and the index i means the month under consideration. H refers to hours, and DAYS indicates days in the month.

Values of p were calculated by the above expressions and compared to the values shown in Blaney and Criddle (1962). Equations 2.38 and 2.39 yielded values that were usually within 1/2% of the 1962 Blaney-Criddle amounts. Chapter 8 shows that the average error was only 0.08%.

Table 2.4 is composed of values calculated using these equations.

2.3.15 *Extraterrestrial radiation* R_a

Extraterrestrial radiation, R_a, is the solar radiation received on a horizontal surface above the earth's atmosphere and is very useful in predicting ground level solar radiation perpendicular to the sun's rays. Extraterrestrial radiation, R_a , is a function of latitude and date only if radiation from the sun perpendicular to the sun's rays is assumed to be constant. Because the earth's axis is tilted and the distance from the sun to the earth varies with season, the geometry involved is complex but merely represents an integration of the functional relationship between the zenith angle of the sun with an empirical distance correction.

$$R_a = SC\left[\frac{\bar{d}}{d}\right]^2 \int_{sunrise}^{sunset} \cos\theta_z dt \qquad (2.40)$$

Sellers (1965) has a detailed derivation of the following equation.

$$R_a = \frac{SC}{\pi}\left[\frac{\bar{d}}{d}\right]^2\left[\omega_{hs}\left\{\frac{\pi}{180}\right\} \sin L \sin\delta + \cos L \cos\delta \sin\omega_{hs}\right] \qquad (2.41)$$

where R_a is *extraterrestrial solar radiation* in MJ m^{-2} day^{-1}, d is the distance to the sun at any time, $\bar{d}$ is the yearly average distance between the sun and the earth, and all angles are defined above. SC is the solar constant, MJ m^{-2} day^{-1}.

Occasionally, alternate algorithms are needed because not all spreadsheets and computer languages have a complete set of trigonometric functions available for use. See Chapter 8 for an alternate algorithm for estimating the sun-earth distance correction factor.

Table 2.4
Values of the monthly percent of daytime on a yearly basis
(positive latitudes are north and negative are south)

lat	Jan	Feb	Mar	April	May	June	July	Aug	Sept	Oct	Nov	Dec
60	4.54	5.62	8.13	8.56	12.06	12.81	12.73	11.02	8.69	6.98	4.90	3.97
50	5.93	6.29	8.25	8.45	10.85	11.15	11.24	10.19	8.54	7.46	6.04	5.62
40	6.73	6.71	8.32	8.38	10.12	10.22	10.38	9.68	8.44	7.77	6.71	6.53
30	7.30	7.01	8.37	8.33	9.60	9.58	9.78	9.30	8.37	8.00	7.19	7.16
20	7.74	7.26	8.42	8.29	9.19	9.07	9.30	9.00	8.32	8.18	7.58	7.66
10	8.13	7.47	8.46	8.25	8.83	8.63	8.88	8.74	8.27	8.34	7.91	8.09
0	8.49	7.67	8.49	8.22	8.49	8.22	8.49	8.49	8.22	8.49	8.22	8.49
-10	8.85	7.87	8.53	8.19	8.16	7.81	8.11	8.25	8.17	8.64	8.53	8.89
-20	9.24	8.08	8.57	8.15	7.80	7.37	7.69	7.99	8.12	8.80	8.86	9.32
-30	9.68	8.32	8.61	8.11	7.39	6.87	7.22	7.69	8.07	8.98	9.24	9.82
-40	10.23	8.62	8.66	8.06	6.88	6.24	6.62	7.32	8.00	9.21	9.71	10.44
-50	11.01	9.03	8.74	7.99	6.18	5.34	5.79	6.82	7.91	9.51	10.36	11.32

2.3.16 *Clear day solar radiation* R_{so}

Clear day or cloudless day radiation, R_{so}, is the solar energy received on a horizontal surface when no clouds attenuate the radiation. Atmospheric constituents, such as dust or water not in the form of clouds do, however, diminish the radiation. R_{so} is best determined by taking the outer envelope of a plot of several seasons solar radiation measurements against time. Allen et al. (1989) indicate that for Utah and Idaho conditions the following relationship holds:

$$R_{so} = 0.75 R_a \tag{2.42}$$

Table 2.5
Extraterrestrial solar radiation
estimated by Equation 2.41
for various latitudes and months
MJ m^{-2} day^{-1}

item	Jan	Feb	Mar	April	May	June	July	Aug	Sept	Oct	Nov	Dec
60	3.3	8.2	16.9	27.8	36.6	41.0	39.2	31.9	21.6	11.8	4.8	2.3
50	8.9	14.3	22.5	31.7	38.5	41.6	40.3	34.9	26.5	17.7	10.7	7.7
40	15.1	20.3	27.5	34.8	39.7	41.8	40.9	37.1	30.7	23.3	16.8	13.8
30	21.1	25.8	31.6	37.0	40.0	41.1	40.6	38.3	33.9	28.2	22.6	19.9
20	26.8	30.5	34.8	38.0	39.3	39.5	39.3	38.4	36.1	32.3	28.0	25.8
10	31.9	34.5	36.9	37.9	37.5	37.0	37.1	37.5	37.2	35.4	32.7	31.1
0	36.2	37.4	37.9	36.8	34.8	33.5	33.8	35.6	37.2	37.5	36.5	35.7
-10	39.5	39.3	37.8	34.5	31.1	29.2	29.8	32.6	36.1	38.5	39.3	39.4
-20	41.9	40.1	36.5	31.2	26.6	24.2	25.0	28.7	33.8	38.3	41.1	42.1
-30	43.1	39.7	34.1	27.0	21.4	18.6	19.6	24.0	30.6	37.0	41.7	43.8
-40	43.3	38.2	30.7	22.1	15.8	12.9	13.9	18.7	26.4	34.6	41.3	44.4
-50	42.6	35.7	26.3	16.5	10.0	7.2	8.1	13.0	21.4	31.2	40.0	44.3

2.3.17 *Tabular estimates of clear day solar radiation* R_{so}

Tabular values of worldwide estimates of clear day solar radiation estimates are shown in Table 2.6 which was adapted from Budyko (1963) by the *Irrigation Water Requirements Committee of ASCE* (Jensen et al., 1990).

2.3.18 *ASHRAE clear day radiation model* R_{so} (*instantaneous*)

There are many clear day solar radiation models (Bird and Hulstrom, 1982 and Iqbal, 1983). These range in complexity from simplified to complex computer models. The following simplified models from *The American Society of Heating, Refrigerating and Air Conditioning Engineers*, ASHRAE, were selected arbitrarily to illustrate general approaches and principles. In addition, workers in fields referring to evaporation need to be aware of appropriate literature existing in other disciplines. ASHRAE certainly receives widespread recognition.

Table 2.6
Clear day solar radiation from Budyko (1963)
and ASCE (Jensen et al., 1990)
for various latitudes and months
MJ m^{-2} day^{-1}

Lat	Jan	Feb	Mar	April	May	June	July	Aug	Sept	Oct	Nov	Dec
60	2.5	6.0	13.8	22.3	29.0	32.0	29.9	23.3	15.8	8.5	3.6	1.6
50	6.7	11.4	18.6	25.8	31.0	33.1	31.5	26.7	20.1	13.5	8.1	5.4
40	12.3	17.0	22.9	28.3	32.1	33.5	32.7	29.2	23.7	18.4	13.5	10.8
30	17.5	21.7	26.0	29.9	32.1	33.2	32.7	30.4	26.7	22.5	18.3	16.0
20	21.7	25.0	28.2	30.1	31.4	31.8	31.5	30.1	28.5	25.8	22.5	20.5
10	25.5	27.6	29.4	29.6	29.6	29.3	29.4	28.8	29.6	28.1	25.8	24.4
0	28.2	29.2	30.0	28.5	26.9	26.3	26.7	27.8	29.6	29.6	28.5	26.8
-10	30.7	30.4	29.4	26.8	24.7	22.7	22.7	25.0	27.6	29.9	30.4	30.7
-20	32.4	30.7	27.9	24.0	20.5	18.7	19.3	22.1	25.5	29.3	31.5	33.0
-30	33.4	30.4	26.0	20.8	16.6	14.8	15.5	18.8	22.9	28.1	32.1	34.3
-40	33.5	28.8	22.9	17.0	12.1	10.2	11.3	14.8	19.3	26.1	31.8	35.2
-50	33.0	26.4	19.3	12.9	7.5	5.3	6.4	10.6	16.2	23.2	30.4	34.9

Iqbal (1983) describes a simplified model for estimating the rate of clear day solar radiation described as the *ASHRAE* algorithm. This method makes use of a common mathematical expression $\exp(-KX)$ where K and X are parameters representing variables influencing radiant heat transfer such as atmospheric turbidity or elevation. This formulation is substantially what is known as Lambert's, Beer's or Bouguer's law which concerns the passage of light rays through the atmosphere (Iqbal, 1983). Many of the methods of estimating the influence of variables, such as dust or water vapor, on the passage of shortwave radiation through the atmosphere make use of this mathematical formulation.

There appear to be two sets of equations for estimating the rate of clear day solar radiation. The first described as Model A here is discussed by Iqbal (1983). The procedure is described by Bird and Hulstrom (1981) as the ASHRAE model. However, more recent publications of the American Society of Heating, Refrigerating and Air Conditioning Engineers, ASHRAE, (1987) present different algorithms which are referred to as Model B.

Clear day solar radiation, R_{so}, is known as *global* radiation meaning that it is a sum of direct shortwave radiation from the sun and diffuse solar radiation which is scattered by constituents in the atmosphere. This may be stated as follows:

$$R_{so} = R_{so(d)} + R_{so(s)} \tag{2.43}$$

where the subscripts *d* and *s* indicate direct radiation and diffuse or radiation redirected by scattering caused by atmospheric constituents. Direct clear day solar radiation is the larger of the two in magnitude. Both components of R_{so} are estimated as a fraction of *apparent extraterrestrial irradiance.*

Model A

The direct portion of R_{so} is estimated by the following:

$$R_{so(d)} = \cos\theta_z\, I_n = \cos\theta_z\, (CN)A \cdot \exp\left[-B\left\{\frac{p}{p_o}\right\}\sec\theta_z\right] \tag{2.44}$$

where A is referred to as *apparent extraterrestrial irradiance* in W m^{-2} or MJ m^{-2} day^{-1}. A as used here is a function of date only and may be assumed to be solar radiation as a maximum only and not related to direction. Latitude is accounted for by Equation 2.28. B is a fitting constant that by definition recognizes the *air mass,* B is shown in Table 2.7. The ratio *p* over p_o has the effect of including an elevation correction. The altitude correction has the effect of increasing R_{so} with increasing elevation. Becker and Boyd (1957) show a general increase of solar radiation of approximately 17% with an elevation of 2000 meters above sea level. θ_z is the zenith angle of the sun previously discussed in Section 2.3.7.

CN is defined as the *clearness number*, CN, which is a dimensionless ratio as follows:

$$CN = \frac{R_{so}(\text{mean site water vapor})}{R_{so}(\text{standard water vapor})} \tag{2.45}$$

where the clearness number is clearly intended to recognize a non-typical water distribution in the air. Iqbal (1983) uses CN = 1 based on the assumption that the visibility is 160 km. Values of CN may be extracted from references cited.

The estimation of the diffuse component of R_{so} is simply a matter of multiplying I_n by an empirical ratio.

$$R_{so(s)} = C\ I_n \tag{2.46}$$

where C is an empirical dimensionless constant or ratio (data for Table 2.7 from Iqbal, 1983 and various publications of ASHRAE).

Model B

$$R_{so(d)} = \cos\theta_z\ I_{dn} = \cos\theta_z\ A/\exp(B/\sin\beta) \tag{2.47}$$

where $\beta = 90° -$ the incident angle θ_z.

$$R_{so(s)} = C\ I_{dn} \tag{2.48}$$

where C is a dimensionless ratio. Apparently the same numerical values of the constants A, B and C are used as in the previous model and the values appear in Table 2.7. A, B and C only differ in units between references.

2.3.19 *Heermann et al. daily estimates* R_{so}

Heermann et al. (1985) made an investigation of solar radiation data from 10 locations ranging all over the United States prompted by the prediction of ET for irrigation scheduling. The interest in clear day solar radiation is because of its use in predicting net radiation. Two sets of empirical equations resulted. The first involved the use of an exponential expression as seen in the ASHRAE models above. The second involves a cosine function.

Table 2.7
Values of A, B and C
from Iqbal (1983) and
various publications of ASHRAE

	ASHRAE			ASHRAE rev. Iqbal(1983)		
date	A	B	C	A	B	C
Jan	1,230	0.142	0.058	1,202	0.141	0.103
Feb	1,215	0.144	0.060	1,187	0.142	0.104
Mar	1,186	0.156	0.071	1,164	0.149	0.109
April	1,136	0.180	0.097	1,130	0.164	0.120
May	1,104	0.196	0.121	1,106	0.177	0.130
June	1,088	0.205	0.134	1,092	0.185	0.137
July	1,085	0.207	0.136	1,093	0.186	0.138
Aug	1,107	0.201	0.122	1,107	0.182	0.134
Sept	1,151	0.177	0.092	1,136	0.165	0.121
Oct	1,192	0.160	0.073	1,136	0.152	0.111
Nov	1,221	0.149	0.063	1,190	0.144	0.106
Dec	1,233	0.142	0.057	1,204	0.141	0.103

A $W\,m^{-2}$
B air $mass^{-1}$
C dimensionless

values are for the 21st of each month

exponential model

$$R_{so} = \dot{A}\ \exp-[(D-\dot{C})/\dot{B}]^2 \qquad (2.49)$$

where $\dot{A}$ is defined as the peak value of radiation, $\dot{C}$ is the calendar day number of the day when the maximum solar radiation occurs, D is the calendar day number of the day under consideration and $\dot{B}$ is said to determine the width of the curve. In theory the maximum solar radiation should occur at day 172 but data showed that this actually occurred at day 170 which was used in their examples.

$$\dot{A} = 31.25 + 0.001113\ z \qquad (2.50)$$

$$\dot{B} = 270 - 3.008\,L \qquad (2.51)$$

where L is latitude in degrees and z is elevation above sea level in m.

cosine model

$$R_{so} = A' + B' \cos(2\pi\,D/365 - C') \qquad (2.52)$$

where

$$A' = 31.54 - 0.2734\,L + 0.0007813\,z \qquad (2.53)$$

$$B' = -0.2986 + 0.2678\,L + 0.0004102\,z \qquad (2.54)$$

where L and z have been previously defined. C' is defined as a phase constant related to the longest day of the year, D = 172. Heermann et al. (1985) use a value of 2.92 for C'. The argument of the cosine function is in radians. The prime superscripts on A, B and C indicate that these constants refer to the cosine model.

They state that the maximum error should be within 5%. They also indicate that the exponential form of the equation is better in spring and fall but is not suitable for winter estimates. A latitude range of 25°N to 47°N is said to result in the best estimates. The equations are said to fail at < 13°N and at > 65°N.

Tables 2.8 to 2.10 were produced using a spreadsheet and show estimates of clear day solar radiation using both the exponential and cosine models of Heermann et al. (1985) and for two elevations using their cosine model. Just values for the northern hemisphere are shown because only US data was used for calibration by Heermann et al. (1985). An attempt was made to limit latitudes to those recommended by the authors.

Table 2.8
Clear day solar radiation R_{so} exponential model
Heermann et al. (1985)
1830 m elevation, MJ m^{-2} day^{-1}
only northern hemisphere is shown

lat	Jan	Feb	Mar	April	May	June	July	Aug	Sept	Oct	Nov	Dec
50	6.2	11.2	17.5	24.8	30.6	33.3	31.8	26.5	19.4	12.6	7.1	3.5
40	11.4	16.6	22.1	27.6	31.5	33.3	32.3	28.8	23.6	17.9	12.4	8.0
30	15.8	20.5	25.1	29.2	32.1	33.3	32.6	30.1	26.2	21.7	16.8	12.4
20	19.3	23.4	27.0	30.3	32.4	33.3	32.8	30.9	27.9	24.3	20.1	16.1

Table 2.9
Clear day solar radiation R_{so} cosine model
Heermann et al. (1985)
1830 m elevation, MJ m^{-2} day^{-1}
only northern hemisphere is shown

lat	Jan	Feb	Mar	April	May	June	July	Aug	Sept	Oct	Nov	Dec
40	12.1	16.0	21.2	27.0	31.3	33.2	32.1	28.2	22.6	17.0	12.6	10.9
30	17.2	20.2	24.1	28.5	31.8	33.2	32.4	29.4	25.2	20.9	17.6	16.3
20	22.3	24.3	27.1	30.1	32.3	33.3	32.7	30.7	27.8	24.9	22.6	21.7
10	27.5	28.5	30.0	31.6	32.8	33.4	33.0	32.0	30.4	28.8	27.6	27.1
0	32.6	32.7	32.9	33.2	33.3	33.4	33.4	33.2	33.0	32.8	32.6	32.5

Table 2.10
Clear day solar radiation R_{so} cosine model
Heermann et al. (1985)
sea level elevation, MJ m^{-2} day^{-1}
only northern hemisphere is shown

lat	Jan	Feb	Mar	April	May	June	July	Aug	Sept	Oct	Nov	Dec
50	6.3	10.8	16.9	23.7	28.7	30.9	29.6	25.1	18.5	12.0	6.8	4.8
40	11.4	14.9	19.8	25.2	29.2	31.0	30.0	26.3	21.1	15.9	11.8	10.2
30	16.5	19.1	22.8	26.8	29.7	31.1	30.3	27.6	23.7	19.9	16.8	15.6
20	21.6	23.3	25.7	28.3	30.3	31.1	30.6	28.9	26.3	23.8	21.8	21.0
10	26.7	27.5	28.6	29.9	30.8	31.2	30.9	30.1	28.9	27.7	26.8	26.4
0	31.8	31.7	31.6	31.4	31.3	31.2	31.3	31.4	31.5	31.7	31.8	31.8

2.4 Solar radiation R_s

Solar radiation striking the earth's surface is obviously attenuated by atmospheric constituents, especially by clouds. Estimation methods of solar radiation at the earth's surface have been based upon either extraterrestrial solar radiation or clear day solar radiation and some indication of cloud cover. Estimated values of extraterrestrial solar radiation and clear day solar radiation are readily available and are shown in Tables 2.5 and 2.6.

Percent, or the ratio of hours of bright sunshine to the possible hours of bright sunshine, S, was measured at many weather bureau stations using either electronic sensors that detected sunshine of a threshold intensity or using magnifying glasses which actually burned a trace on a chart. S is equal to

$$S = \frac{n}{N} \tag{2.55}$$

where n is the measured number of hours of bright sunshine, and N is the possible number of hours of bright sunshine at a location and time. Presumably such measurements are being made less now because direct measurements of solar radiation using automated climate stations are easier to make.

Apparently one of the first regression equations relating solar radiation received on a horizontal surface at a particular time and place to clear day solar radiation and the sunshine ratio was that proposed by Angstrom (1924).

$$R_s = R_{so}(a + b \cdot S) \tag{2.56}$$

where R_s is the *solar radiation*, diffuse and direct, received at the earth's surface on a horizontal surface, R_{so} is the *clear day solar radiation* received on a horizontal surface, S is the *ratio of the actual duration of sunshine to the maximum possible duration of sunshine* for the day, and a and b are empirical constants (actually in Angstrom's presentation this was stated as $b = 1 - a$). Angstrom proposed specific constants of 0.25 and 0.75. Later research has refined the constants, but the basic regression equation is known as an *Angstrom equation*.

Penman (1948) suggested estimating ground level solar radiation using sunshine and extraterrestrial radiation R_a data as:

$$R_s = R_a(a + b \cdot S) \tag{2.57}$$

Doorenbos and Pruitt (1977) recommended values of 0.25 and 0.5 for regression constants for general use. They also show an extensive table of values for more specific locations and uses. Note that $(a+b)$ are always less than 1 because less than 100% of the extraterrestrial solar radiation would be expected to be received at the earth's surface over a daily or monthly period.

At many airports, visual estimates of cloud cover are recorded at observation times. The measurement is known as *fractional cloud cover*, C, and is obviously subjective in character. In the US and many other countries such data is often more available than either solar radiation or percent sunshine data. C is zero for a cloudless sky and has a value of unity for complete cloud cover. Jensen et al. (1990) state that solar radiation estimates made using C are not as reliable as those using S because of the subjective nature of C. Brutsaert (1982) indicates that in principle C + S should equal unity but that such a relationship should only be used when no other data are available.

Correlations of solar radiation and C are available using either R_{so} or R_a. An example of a correlation follows, Kimball (1928).

$$R_s = R_{so}(0.29 + 0.71(1 - C)) \tag{2.58}$$

The authors question the previous equation because R_s only equals 0.29 times R_{so} when C = 1. The general consensus is that estimates of R_s made using S are more reliable than those made using C.

A method of making daily estimates of solar radiation involving rainy days and humidity is shown by Reddy (1971).

2.5 Net radiation_R_n

Net radiation is much more difficult to measure than is solar radiation because of sensor difficulties and because of variations that occur due to surface conditions. Net radiation is usually measured only as part of research projects involving ET. Thus, it is common to estimate net radiation using one of several models based upon solar radiation and parameters such as air temperature and humidity.

2.5.1 *Net radiation estimation* R_n

Net radiation, R_n, is the difference between incoming and outgoing radiation of both short wavelengths from the sun and long wavelengths from the earth and also from clouds and other material in the sky.

$$R_n = S\downarrow - S\uparrow + L\downarrow - L\uparrow \tag{2.59}$$

$$R_n = R_s(1 - \alpha) + L\downarrow - L\uparrow \tag{2.60}$$

$$R_n = R_s(1 - \alpha) + R_{nl} \tag{2.61}$$

$$R_n = R_{ns} + R_{nl} \tag{2.62}$$

where S and L are respectively short and long wave radiation. Fluxes are defined as positive when they are toward the ground surface. The terms R_n, R_{nl}, L, and S are radiant energy in MJ m^{-2} day^{-1}, and α is the dimensionless ratio called albedo. Table 2.2 shows typical values of α for representative surfaces. For reference crops, α is often assumed to have a value of 0.23 for alfalfa (Jensen et al., 1990) and 0.25 for grass (Doorenbos and Pruitt, 1977). Wright (1982) presents a model where α varies throughout the growing season. R_{ns}, is called *net shortwave radiation* and R_{nl}, is a combined term known as *net outgoing long wave radiation*. The long wave radiation terms involving $L\uparrow$ and $L\downarrow$, as shown in Equations 2.59 and 2.60, are solved in a variety of ways. Weiss (1982) retained the identity of each term using a sky radiation model developed by Brutsaert (1975) and incorporated a cloudiness correction of Linacre (1967b) to estimate R_n.

The Stefan-Boltzmann equation for the determination of radiant heat transfer is used to determine the long wave radiation, R_{nl}. Two basic approaches are in common usage. The first, known as a *Brunt* equation, uses ground level air temperatures and calculates a combined ε'_{bulk} for both the air and the ground level soil and plant surfaces. The second approach retains the identity of both the atmosphere and the ground level surfaces. Shelter temperatures are used as well in this variation. This approach is called a *component* approach here.

2.5.2 *Net long wave radiation using Brunt equation*

Because of common usage the temperature term in the Stefan-Boltzmann equation has been used as an average of the maximum air temperature to the fourth power and the minimum air temperature to the fourth power.

$$R_{nl} = L\uparrow - L\downarrow = \varepsilon_{bulk}\sigma\left[\frac{T^4_{k(\max)} + T^4_{k(\min)}}{2}\right] \tag{2.63}$$

where σ has been defined previously. ε_{bulk} is a combined emissivity for both the sky and the ground surface (this term is also known as *net emittance*), $T_{k(max)}$ is the *daily maximum air temperature* or period average maximum air temperature in K, and $T_{k(min)}$ is the *daily minimum air temperature* or period average minimum air temperature in K. σ is first discussed in Section 2.3.

Brunt (1932) proposed an empirical formula for determining the emissivity, ε, for the estimation of *downward radiation from the sky* using a linear regression on the square root of the vapor pressure of the air at the location.

$$\varepsilon_{bulk} = a + b\sqrt{e_a} \tag{2.64}$$

where a and b are simply regression coefficients.

2.5.3 *Constant coefficients for Brunt equation*

Jensen et al. (1974, 1990), Burman et al. (1983) and others show the following for the estimation of ε_{bulk}.

$$CR = \frac{R_s}{R_{so}} \tag{2.65}$$

$$\varepsilon_{bulk} = -[a(CR) + b][a_1 + b_1\sqrt{e_a}] \qquad (2.66)$$

$$R_n = R_s(1-\alpha) + R_{nl} = R_s(1-\alpha) + \varepsilon_{bulk}\sigma\left[\frac{T^4_{k(\max)} + T^4_{k(\min)}}{2}\right] \qquad (2.67)$$

where a and b are regression coefficients for a cloudiness correction, and a_1 and b_1 are regression coefficients for a correction due to water in the air, usually at shelter height. CR is dimensionless and e_a is the *actual vapor pressure of the air* kPa. Here e_a is in kPa, R_{so} is *clear day solar radiation* in MJ m^{-2} day^{-1}, and R_s is actual *solar radiation* in MJ m^{-2} day^{-1}.

An alternate method of estimating R_{nl} when humidity data is not available was developed by Idso and Jackson (1969).

$$\varepsilon_{bulk} = -[a(CR) + b]\{-0.02 + 0.261\exp[-7.77 \cdot 10^{-4}(273 - T_k)^2]\} \qquad (2.68)$$

where T_k is the *average daily temperature* in K, and the other variables are defined above.

Table 2.11 shows values of *a, b,* a_1 *and* b_1 for various locations. These values are constant throughout the growing season and for any location once a selection of values has been made.

2.5.4 *Variable coefficients for Brunt equation*

Wright (1982) in order to improve the fit between measured and estimated ET introduced a model for the estimation of R_n using variable coefficients in calculations. The coefficients *a, b,* a_1 *and* b_1 have the same meaning as those in the previous section. Data for the model came from Kimberly, Idaho, but Jensen et al. (1990) report that estimates of R_n using the model from a variety of locations fit measurements quite well. Wright's system follows:

$$\alpha = 0.29 + 0.06\sin[30(m + 0.0333\,N + 2.25)] \qquad (2.69)$$

$$a_1 = 0.26 + 0.1\exp-[0.0154(30m + N - 207)]^2 \qquad (2.70)$$

$$b_1 = -0.139 \qquad \textit{always} \qquad (2.71)$$

$$\begin{aligned} &\textit{if} \quad R_s/R_{so} > 0.7 \\ &a = 1.126 \qquad b = -0.07 \end{aligned} \qquad (2.72)$$

$$if \quad R_s/R_{so} \le 0.7$$

$$a = 1.017 \qquad b = -0.06 \tag{2.73}$$

$$R_n = R_s(1-\alpha) - [a(CR)+b][a_1 + b_1\sqrt{e_a}]\sigma\left[\frac{T^4_{k(\max)} + T^4_{k(\min)}}{2}\right] \tag{2.74}$$

where R_s and R_{so} have been previously defined, m is the *month number*, and N is the *day number* in the month. The parameter m is equal to 1 for January in the northern hemisphere; m is offset by 6 months in the southern hemisphere.

2.5.5 *Net long wave radiation R_{nl} using component approach*

When using the component equation to estimate R_{nl} (Weiss, 1982), T_{ka} is the average of the daily maximum and minimum temperatures in K for use in Stefan's equation. By custom both the maximum and minimum temperatures are used individually in Stefan's equation by raising each temperature to the fourth power and the result is averaged in the common use of the Brunt equation.

The component procedure of estimating R_{nl} has the advantage of making it possible for the estimator to determine separate values of the emissivity ε for both the sky and the ground surface. This should make it possible to make allowances for different crops and ground surface conditions. The added flexibility may permit more control over estimates and possibly improve estimates.

The procedure has two steps.

1. Estimate the long wave radiation components using Stefan's equation, Section 2.3 Radiation. In contrast to the combined parameter approach, the component method uses the average daily air shelter temperature (Weiss, 1982).

a. Estimate the long wave radiation leaving the earth's surface. This radiation is received by the atmosphere or lost into space. See Equation 2.76.

b. Estimate the long wave radiation received at the ground surface from the sky. The long wave radiation from the sky is almost always much less than the long wave radiation leaving the earth's surface. See Equations 2.77-2.79.

2. Apply an empirical correction for cloudiness. See equation 2.80.

Table 2.11
Values of regression coefficients for estimating
combined surface and sky albedo (Jensen et al., 1990)

location	a	b	a_l	b_l	source/application
Davis, CA	1.35	-0.35	0.35	-0.145	Pruitt 73; Goss & Brooks 56
So. Idaho	1.22	-0.18	0.325	-0.139	Wright & Jensen 72
England			0.47	-0.206	Monteith & Szeicz 62
England			0.44	-0.253	Penman, 48
Australia			0.35	-0.133	Fitz. & Stern 65
General	1.2	-0.2	0.39	-0.158	arid areas
General	1	0			humid & semi humid

Net long wave radiation is estimated as follows:

$$R_{nl} = C_{clouds}(L\downarrow + L\uparrow) \tag{2.75}$$

where C_{clouds} is a dimensionless ratio calculated using Equation 2.80.

The outward long wave radiation component is estimated using a simple direct application of Stefan's law.

$$L\uparrow = \varepsilon_{surface}\sigma T_k^4 \tag{2.76}$$

where $\varepsilon_{surface}$ is the *emissivity* of the plant, ground or combined surface, σ is the *Stefan-Boltzmann constant*, and T_k is the *average daily air temperature* in K. As an expedient, the shelter height air temperature is used as the ground surface temperature because ground surface temperature is only measured under research conditions. The following table is a summary of values of $\varepsilon_{surface}$ for different crops and surfaces.

Table 2.12
Common values of emissivity

surface	gen avg	low	high	remarks
snow	0.98	0.97	0.99	varies with age
bare soil	0.97	0.95	0.98	
green vegetation	0.97	0.96	0.98	
water	0.96	0.96	0.96	
bricks	0.72	0.45	0.93	depends upon kind
aluminum shapes	0.17	0.04	0.31	

The estimation of the term representing long wave radiation from the sky R_{nl} is difficult because the only data available for routine estimates of ET are shelter air temperature and perhaps shelter vapor pressure. The temperature distribution of the atmosphere as well as the distribution of water and other constituents is unknown. The use of tabular values of ε is inadequate. Brutsaert (1975) proposed the following empirical function for the estimation of ε_{sky}.

$$L\downarrow = \varepsilon_{sky}\sigma T_k^4 \tag{2.77}$$

$$\varepsilon_{sky} = 1.24\left[10\frac{e_a}{T_k}\right]^{\frac{1}{7}} \tag{2.78}$$

$$\text{if} \quad C_{clouds} > 0.7 \qquad \varepsilon_{sky} = 0.97\varepsilon_{sky} \tag{2.79}$$

where e_a is average *vapor pressure* in kPa, and T_k is the average *daily air temperature* in K. Weiss (1982) found the equations to be reliable and recommended them.

The cloudiness correction recommended by Weiss (1982) is an adaptation of one proposed by Linacre (1968). Weiss (1982) recommended different coefficients than Linacre (1968). The correction used by Weiss follows:

$$C_{clouds} = (0.4 + 0.6\, R_s/R_{so}) \tag{2.80}$$

where all variables have been identified previously.

2.6 Soil heat flux

Soil heat flux, G, is small compared to R_s, R_n, ET, and other terms of the energy balance and, therefore, is often neglected in making estimates of ET. Because of the relatively small magnitude of G, its influence on final estimates is small. The principal reasons for including G are its inclusion recognizes a realistic phenomenon, and any simple model of G is easy to include in making computerized estimates of ET. Brutsaert (1982) discusses the differential equations defining soil heat flux. Because air temperature is a trigonometric function, the more theoretically based models involve periodic functions.

The following models are widely recommended but are approximations of a very complex process. The following algorithms are based on the idea that soil temperature follows air temperature and approximations of the effective root zone and the heat capacity of the soil. Constant values of soil heat capacity are utilized when the soil heat storage capacity varies constantly with the soil water present. These models are easy to include in computer programs and utilize only a minimal amount of computer time.

2.6.1 *Daily model*

Kincaid and Heermann (1974) and Wright (1982) recommend the following algorithm which is based on broad assumptions of the root zone depth, the soil heat capacity, and the notion that soil temperature lags air temperature.

$$G = 0.3768[\overline{T}_D - (\overline{T}_{D-1} + \overline{T}_{D-2} + \overline{T}_{D-3})/3] \tag{2.81}$$

where G is daily *soil heat flux* in MJ m^{-2} d^{-1}, D indicates the *calendar day number* under consideration, and T values are daily average temperatures in °C. The constant 0.3768 varies with the units of G and the air temperatures T. The sign convention results in G being negative in the spring when the soil is warming. At that time G will slightly decrease estimated ET.

2.6.2 *Ten day to monthly model*

A simplified model for G recommended for periods of from 10 to 30 days is presented by Jensen et al. (1990). The formula is based on a soil with 50% solids, negligible organic material, and a volumetric soil water capacity of 27%. A 2 meter effective soil depth is assumed.

$$G = 4.19[\overline{T}_{J+1} - \overline{T}_{J-1}]/\Delta time \tag{2.82}$$

where G is MJ m^{-2} d^{-1}, and the air temperatures T are average temperatures in °C, and the subscripts J refer to the periods selected. The time period is in days. For example, when using monthly data $\Delta time$ has a value of 60. The constant 4.19 is based on the assumptions shown above and unit considerations.

Chapter 3

SOIL PARAMETERS

Soil in the root zone provides structural support, a reservoir for water and a suitable chemical, biological, and thermodynamic environment for healthy growing plants. The suitable chemical and thermodynamic environment is of unusual importance. The chemical and biological environments will not be discussed here because this subject matter is not appropriate for this volume. Major discussion about the role of the soil profile in providing water for plant maintenance and growth will appear in this chapter. Dry soil obviously inhibits plant growth and transpiration and in extreme cases even kills plants. Excessive soil water will have an adverse effect on most commercial crops. Siemer and Rumberg (1975) state that soil water conditions and fertility levels determine the kinds of plants growing in high altitude conditions where normal cultivation is not practiced. Infiltration of water into the root zone is not discussed herein because the subject is discussed in material dealing with soil physics, irrigation, or hydrology. The thermodynamic environment is not discussed in this chapter. Soil heat flux is a component in the energy balance equation as it applies to ET. Two simplified models for predicting *soil heat flux*, G, are described in chapter 2.

3.1 Soil water fundamentals

The retention and movement of water in the soil follow many of the fundamental laws of physics and mathematics. Therefore, it is possible to relate much of the material to theory. The complexity of the theories is formidable even for a hypothetically uniform soil. However, the natural variability of soil makes the task of understanding complex and difficult. The hydrologist needs a grasp of soil water fundamentals to estimate correctly hydrologic events.

3.1.1 *Energy status of soil water*

Soil water potential, ψ, is defined as the ability to do work. This involves the work required to move a quantity of pure water from a reference point to conditions prevalent at the point under consideration. This kind of theory is not unusual and should not be difficult for the reader to understand. ψ represents energy. By definition, ψ has the units of (force) x (distance). Even though work or energy units are used, ψ may be described in at least three unit systems (Hillel, 1980). These include:

1. **Energy per unit mass** $Nm\,kg^{-1}$. These units are often used by engineers dealing with fluid mechanics or thermodynamics as $V^2\,m^{-2}$. These units are not normally used in soil water descriptions.

2. **Energy per unit volume** $Nm\,m^{-3}$ which reduces to Pa. These units are discussed in references dealing with the topic *dimensional analysis and similitude*. A variation of these units is commonly used for soil water description. It is very common to use a multiple of the bar which is 100 kPa or the standard atmosphere which is 101.3 kPa in soil water designation.

3. **Energy per unit weight** $Nm\,N^{-1}$ which reduces to the simple length unit of *m*. This unit is commonly used by hydraulic engineers and has been used in soil water description.

The hydrologist is faced with a very confusing problem in interpreting the units used to describe soil water because of the proliferation of units that may be found in the literature. This is confusing because the SI, c.g.s. and English systems are imposed on the above variations. In addition, the units *atmospheres* or *bars* are often used interchangeably in describing soil water reference points even though a US Standard Atmosphere (standard sea level pressure) is defined as a pressure of 101.3 kPa, and a bar is defined as a pressure of 100 kPa. The differences are small from a practical standpoint but are major from a theoretical viewpoint.

Soil water potential may be defined as the sum of four component potentials.

$$\psi = \psi_g + \psi_m + \psi_p + \psi_s \tag{3.1}$$

where *g* stands for *gravity*, *m* represents *matrix*, *p* is *pressure* and *s* indicates *osmotic* force.

3.1.2 *Gravity potential*

The weight of water results in a gravitational force causing downward movement, and thus the water does positive work when moving downward. The gravity potential is useful only in referring to the vertical position of a point. Its inclusion has the objective of relating soil water potential to the vertical distance to a water table.

3.1.3 *External pressure potential*

ψ_p is always positive and may be regarded as being caused by an external force. The hydrostatic potential is the positive pressure encountered below a water table. It is easily calculated when the vertical distance from a point of known pressure to a point of unknown pressure is known.

The term ψ_p may be calculated by the hydrostatic equation which follows:

$$\psi_p = P_w = (\rho_w g)z \tag{3.2}$$

where P_w is *hydrostatic pressure* in kPa, ρ_w is water *density* in kg m^{-3}, g is the *acceleration of gravity* in m s^{-2}, and z is the *vertical distance* of the point below the water surface in meters. The hydrostatic pressure is always used as a positive gage pressure.

3.1.4 *Matric potential*

Matric pressure is a negative pressure which results from a combination of capillarity and the affinity of water for soil particles. The affinity of soil particles is known as the phenomena of absorption. Capillarity involves thin passages and curvature of water films. ψ_m is usually shown as a positive number; thus, the minus sign is unfortunately neglected. For dry soil conditions, the value of ψ_m is so large in comparison to ψ_g that only the matric potential is used to describe soil water conditions.

3.1.5 *Osmotic potential*

The osmotic potential ψ_s is caused by the presence of solutes in soil water. The process is particularly important when a membrane is present in the soil. This process is important for the relationship between plant roots and soil; for example, roots imbibing water. Specific discussions of λ_s will not appear in this volume. The reader is referred to the many volumes dealing with soil physics.

3.1.6 *First law of thermodynamics*

Material in this section may be considered to be peripheral in the practical task of estimating ET from climatic data. It is only presented for possible benefit to the reader with a background in fluid mechanics or thermodynamics. A form of the Bernoulli equation has been used to illustrate the energy concept of soil water (Jensen et al., 1990). Perhaps the greatest benefit of the following discussion is to examine the assumptions necessary to apply the conservation of energy principle to soil water, and then to show that the conservation of energy notion has only minor application to soil water.

The classic first law of thermodynamics can be extended to explain partially the energy status of soil water. The first law of thermodynamics is a statement of conservation of energy. In its usual definition, energy is not created or destroyed, only changed from one form to another. Here the assumption is made that nuclear reactions are ignored.

The most common statement of the first law of thermodynamics used by engineers applies to a control volume such as a turbine, valve, or jet engine. A control volume is not applicable to soil water but the conservation of energy in various forms may be useful in explaining the energy status of soil water.

The first law of thermodynamics, or conservation of energy principle, may be stated as follows:

$$\dot{Q} - \dot{W} = \frac{d}{dt}\int_{cv} e\rho d\hat{V} + \int_{cs}\left[\frac{P}{\rho} + e\right]\rho\vec{V} \cdot \overrightarrow{dA} \tag{3.3}$$

$$e = \frac{V^2}{2} + u + gz \tag{3.4}$$

where $\dot{Q}$ is the *rate of heat transfer* in $N \cdot m$ s^{-1}, and $\dot{W}$is *work* in the same units. $\hat{V}$ is *volume* in m^3, ρ is *density* in kg m^{-3}, V is *velocity* in m s^{-1}, and *e* is *energy* in m^2 s^{-2}. The subscript *cv* means the control volume, and the subscript *cs* means control surface. The quantity *u* is *internal energy*. In both cases the units are m^2 s^{-2}. P is *pressure* in kPa, and v is *specific volume* in m^3 kg^{-1}. The nomenclature of Fox and McDonald (1985) is followed.

Assuming that the work term is zero, that heat transfer and internal energy may be ignored, an incompressible fluid, uniform flow and that the unsteady integral may be ignored, the following results:

$$o = \int_{cs} (e + P/\rho)\rho\vec{V} \cdot \overrightarrow{dA} \tag{3.5}$$

Integration with a change in units results in:

$$H = \frac{V^2}{2g} + \frac{P}{\gamma} + z \tag{3.6}$$

where *H* is total energy in m and γ is specific weight in kN m^{-3}.

Equation 3.6 has the same terms as the well known Bernoulli equation but is the result of different assumptions and approach. All terms now have the units of *m*.

Obviously, soil water movement velocities are low; therefore, kinetic energy may be neglected. The kinetic energy term is often neglected even with high velocities when the fluid involved is a gas with a low density compared to water.

The final equation applied to soil water potential ψ_p is:

$$H = \frac{p}{\gamma} + z \tag{3.7}$$

The assumptions needed for even a tenuous application of the conservation of energy principle follow. Most of the assumptions also apply to application of the Bernoulli equation.

1. Ignore heat transfer.
2. Ignore water velocity.
3. Consider water to be incompressible.

Most of the discussion has to do with the vertical distance between a point and a water table. The forces contributing to soil water used in plant growth, such as capillarity and the particle attraction, are usually much greater than any potential resulting from the vertical distance of a point to a water table. Commercial crops are cultivated in fields having a depth to water table of at least two meters. Water retention in a cultivated field ranges from a negative pressure equivalent of a height of 1 meter of water to a negative pressure equivalent of 155 meters of water. The main effect of the water table is confined to a saturated zone located just above the water table which is called the *capillary fringe*.

3.2 Soil water reference points

Soil varies from location to location. The mineral, chemical, and organic material composing soil specifically varies. The chemical, mineral, and biological constituents, as well as their arrangement, influence water retention within soil. A series of specific reference points help define soil characteristics. At times these points have been defined as soil water equilibrium points. The word *reference* is superior to *equilibrium* because the points really do not represent equilibrium because there is no such thing as single valued points for parameters which constantly vary. The following discussion refers to values useful in defining soil water retention in soil.

Recently the nomenclature used to define these points has changed. The following has the objective of pointing out both agreements and disagreements in nomenclature. Both the old and new nomenclature are prominently mentioned in literature dealing with ET and soil water.

3.2.1 *Saturation*

To *saturate* is defined in a standard dictionary (Thorndike and Barnhart, 1974) as *to soak thoroughly or to fill full.* Applied to soils, saturation occurs when the soil pores are full of water. This condition is actually never reached because even saturated soil contains a very small amount of air.

The soil water potential, ψ, at saturation is zero to a positive magnitude at or below the water table. However, a saturated zone exists above a water table and, therefore, the total potential may be slightly negative.

3.2.2 *Drained upper limit or field capacity*

The *drained upper limit*, DUL, or *Field Capacity*, FC, represents the amount of soil water retained by the soil against gravity. The intent of both definitions is the same, but minor differences exist in descriptions of measurements of the parameters. For most practical purposes, the DUL and FC may be regarded as equivalent. The normal maximum root zone is considered, and FC or DUL is determined in depth increments in units such as mm m^{-1}. The determination of FC or DUL is nearly the same for both parameters.

Determination of DUL or FC

Ritchie (1985) describes a method of measuring what he calls the DUL. The soil profile is thoroughly saturated. Then a plastic cover is placed over the soil surface to prevent evaporation. The water content is measured daily using neutron meters or gravimetric means until the water content remains nearly constant. The water content measured at this time is designated as the DUL.

Historically FC has often been determined by field measurements using neutron probes or gravimetric sampling at an arbitrary period of 2 to 3 days after a thorough irrigation. FC is associated with a soil water tension ψ_m of about 1/10 atmosphere for sandy soils and 1/3 atmosphere for clay bearing soils (Hansen et al. 1980).

3.2.3 *Lower limit or wilting point*

The *lower limit*, LL, is used as an index related to soil water available for plant utilization. The LL is apparently the same concept as the *wilting point*, WP, used in water management literature for many years.

The point may be defined as the water content at which plants become wilted and eventually die. The water content is, therefore, dependent on crop and climate. The concept is confusing because a soil profile may contain dry soil below the LL or WP at the ground surface; but because soil water changes with depth, the soil water content near the bottom of the root zone may be much higher.

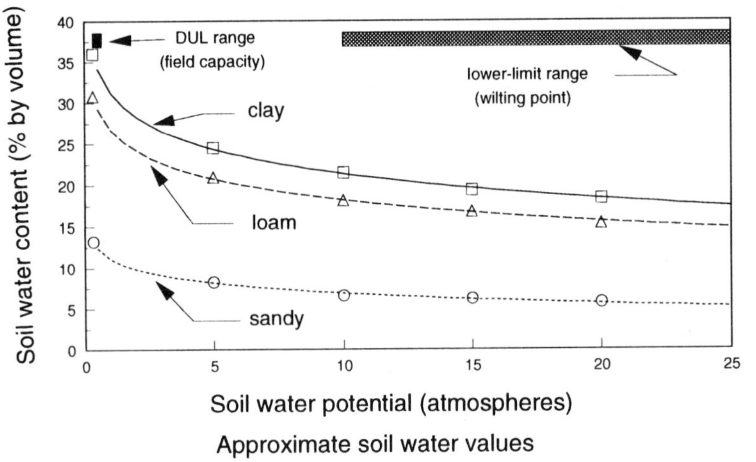

Figure 3.1
Soil water retention related to tension
for different soils

Determination of LL or WP

The LL or WP may be estimated by laboratory measurements. The soil water content at a tension of -15 bars is often taken as the LL or WP for commercial crops, and much higher values are appropriate for range land vegetation. The WP for commercial crops is approximately equal to the FC divided by 2 to 2.4 depending on the silt content of the soil (Hansen et al. 1980). The value, however, changes greatly with atmospheric conditions and the growth stage of the plant.

The LL may be taken as the soil water content at which plants enter permanent wilting and die. Ritchie (1985) recommends the LL be determined by field tests but enumerates the difficulties involved. Problems resulting from natural rainfall preventing the soil from drying and the location of a water table are discussed.

The use of laboratory determination of the amount of soil water retained at a tension of -15 bars has been quite common when determining the wilting point. In early literature the WP was determined by growing plants in pots in greenhouses. An inspection of the general relationship between matric potential and water content reveals that the water content at -10 bars is near the water content at -20 bars. See Figure 3.1.

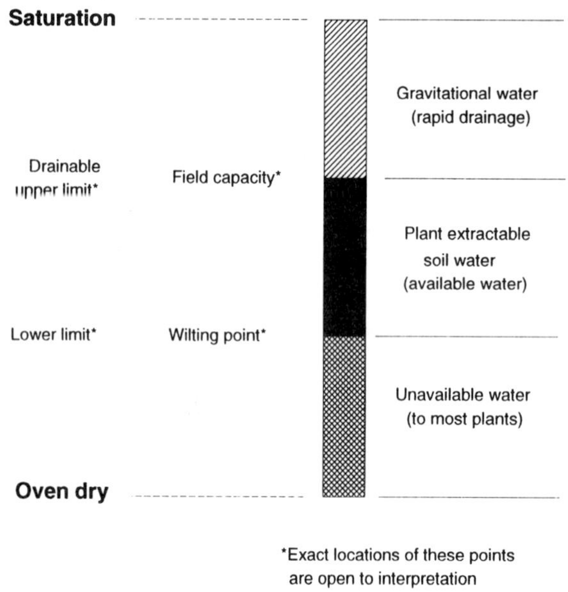

Figure 3.2
Soil water availability classes

3.3 Soil water storage

Soil provides structural support to plants, and, at the same time, provides stored soil water for growth and survival of plants. Soil water holding capacities of soils vary with soil texture, organic material, and many other factors. Numerical values of the volume of water available for plant consumption are useful for a variety of reasons, and examples will be presented in material to follow.

3.3.1 *Plant extractable soil water (available soil water)*

The term *plant extractable soil water*, PESW, is the soil water between DUL and LL. PESW is the same as the term *available soil water*, ASW, and has often been used for irrigation system design and operation. For commercial crops, it becomes the reservoir of water for evapotranspiration ET.

Ratliff et al. (1983) made a comparison of field measured PESW with laboratory estimates where a ψ_m of -0.33 bars was assumed to represent the drainable upper limit DUL, and a ψ_m of -15 bars was associated with the lower limit LL for midwestern soils. They conclude that field measurements of PESW, when available, are preferable to laboratory estimates. The laboratory estimates do, however, provide a reasonable estimate of PESW. The US Department of Agriculture (1970) refers to this as WRD, the *water retention difference*.

Figure 3.2 shows a simple illustration of the availability of soil water to plants. The concepts involved have been very constant, but the exact interpretations of some of the points such as DUL and LL are not completely clear. Nevertheless, specific numerical values are useful in hydrology and specifically in irrigation system design and management.

Figure 3.3 shows how these values vary as a function of general soil type. Even though the individual points change, the plant extractable soil water, PESW, does not change dramatically with soil type.

3.3.2 *Readily available soil water,* RAM

This additional classification of water held in the soil for plant use is the result of the fact that water near the wilting point, a ψ_m of -15 bars or atmospheres, is not available to most commercial crops. Therefore, Hansen et al. (1980) define *readily available soil water*, RAM, as about 75% of available soil water. Depletion of soil water to 75% of available soil water is dubious because irrigation scheduling is usually based on depletions of 50 to 65%, and permissible depletion is known to vary with crop and growth stage.

3.3.3 *Management allowed deficiency* MAD

Management allowed deficiency, MAD, may be defined as the percent of ASW, or its equivalent PESW, in the root zone that corresponds to an acceptable water stress from a management standpoint. This definition was adapted from Merriam et al. (1980). The concept of MAD is useful in irrigation scheduling.

$$MAD = 100 \cdot \left[\frac{RAM}{ASW} \right] \tag{3.8}$$

James (1988) presents a detailed table of MAD values and nominal root zone depths for 39 crops. James shows that the values of MAD vary from 30% to 50% for crops such as carrots and tomatoes to a high of 65% for alfalfa and corn. He indicates that values of MAD change with soil texture.

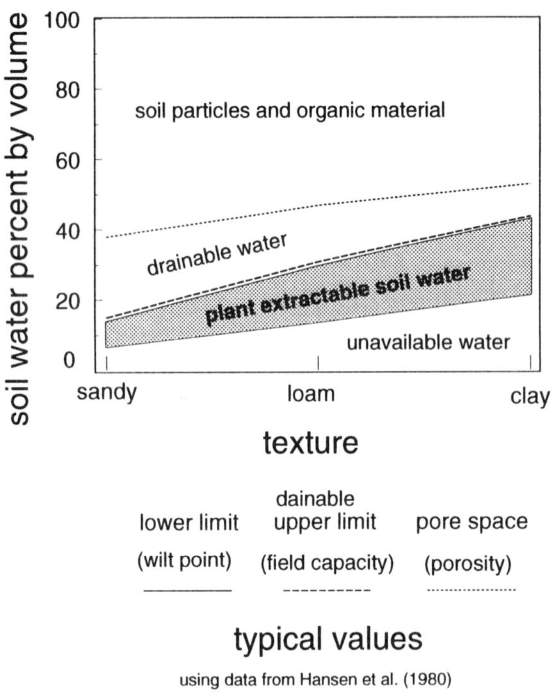

Figure 3.3 Soil texture and soil water availability

3.3.4 *Typical soil water capacities*

Table 3.1 shows typical soil water properties for common soil types encountered in normal commercial cropping. The table was adapted from Rawls et al. (1982). Note that the bulk density values shown are for average conditions. Alexander (1989) shows that the effect of the organic carbon content is so large that empirical equations used for the predictions of bulk density may be developed using organic carbon as the major input, particularly for southern Alaska conditions. Walter et al. (1990) substantiate this effect for a high altitude area in Colorado. Under these conditions, bulk density values may be substantially less than those shown in the table.

Table 3.1
Soil hydrologic data

texture class	effect. porosi %	field cap. DUL %	wilting point LL %	avail. water PESW %	avail. water PESW mm/m	approx. fringe height entry press. mm	sat. cond. mm/hr
sand	41.7	9.1	3.3	5.8	58	73	210.0
loamy sand	40.1	12.5	5.5	7.0	70	87	61.1
sandy loam	41.2	20.7	9.5	11.2	112	147	25.9
loam	43.4	27.0	11.7	15.3	153	112	13.2
silt loam	48.6	33.0	13.3	19.7	197	208	6.8
sandy cl lo	33.0	25.5	14.8	10.7	107	281	4.1
clay loam	39.0	31.8	19.7	12.1	121	259	2.3
silty cl lo	43.2	36.6	20.8	15.8	158	326	1.5
sandy clay	32.1	33.9	23.9	10.0	100	292	1.2
silty clay	42.3	38.7	25.0	13.7	137	342	0.9
clay	38.5	39.6	27.2	12.4	124	373	0.6

after Rawls et al. (1982)

3.3.5 *Field soil water and water table conditions.*

Soil water conditions in a field containing growing plants are quite variable, ranging from complete saturation in many cases such as native meadows in high altitude areas of the US to very dry conditions under desert conditions. The presence of a water table may not be of interest in many irrigated areas, but consideration of a water table and how soil water varies above the water shows the tremendous variability of soil water and wide range of soil water conditions involved in plant growth.

Figure 3.4 shows a saturated zone in the soil profile extending to a short distance above the water table. The water table is defined as the location of zero gage water pressure, and the capillary fringe is a zone of near saturation located above the point of zero water pressure. Therefore, the capillary fringe has a slightly negative pressure, and drains placed in the capillary zone cannot collect water. The capillary zone, however, can provide plant water for ET. The root zone of most plants is an unsaturated zone.

Figure 3.4 also contrasts soil water in a dryland site to that within a typical irrigated field. The primary goal of irrigation is to maintain soil water in the root zone within a range that is optimum for plant growth. Saturated conditions usually occur after a heavy irrigation or rainstorm. Saturated conditions may also be found in soil profiles in early spring after any snowmelt has occurred. The profile is not allowed to dry to a level where plant growth is limited. A common rule of thumb is that irrigation water should be applied when the soil water within the root zone approaches a 50% reduction. The range of soil water under irrigation is shown schematically in Figure 3.4.

The distribution shown in Figure 3.4 for dryland is only one of many that are possible. The sketch shows a possible distribution late in a dry growing season. Soil water just above the water table may be near the DUL (field capacity). However, the same plant may have a surface soil water content significantly below the LL (wilting point). This possibility was discussed by Ritchie (1985) and has been verified by extensive measurements made in Colorado (Walter et al., 1990). The reader should remember that the predominance of plant roots is in the upper layers, and, that even though the bottom of the root zone is near the DUL, the plant may be subject to drouth stress because of dry soil near the surface.

Figure 3.4 shows that plants grow under extremely broad soil water levels. A point not shown in Figure 3.4 is that many plants grow best under saturated conditions. The grass-like plants, known as sedges and rushes, found in high altitude meadows in the US are examples of this phenomenon.

3.4 Soil water influences on plant growth

Soil water conditions have a profound impact on what plants are found at specific locations. Irrigation is used to manage soil water conditions for the optimum production of commercial crops. Detailed discussions of soil water magnitude and its effect on ET and production appear in Chapters 5 and 6 of this volume.

3.5 Soil water movement

Soil water movement is important in considerations of evapotranspiration. Water is relocated in the soil following a rainstorm or an irrigation. Water is moved to the soil surface to supply ET. Plant roots or unsaturated flow through soil particles move the water to the earth's surface.

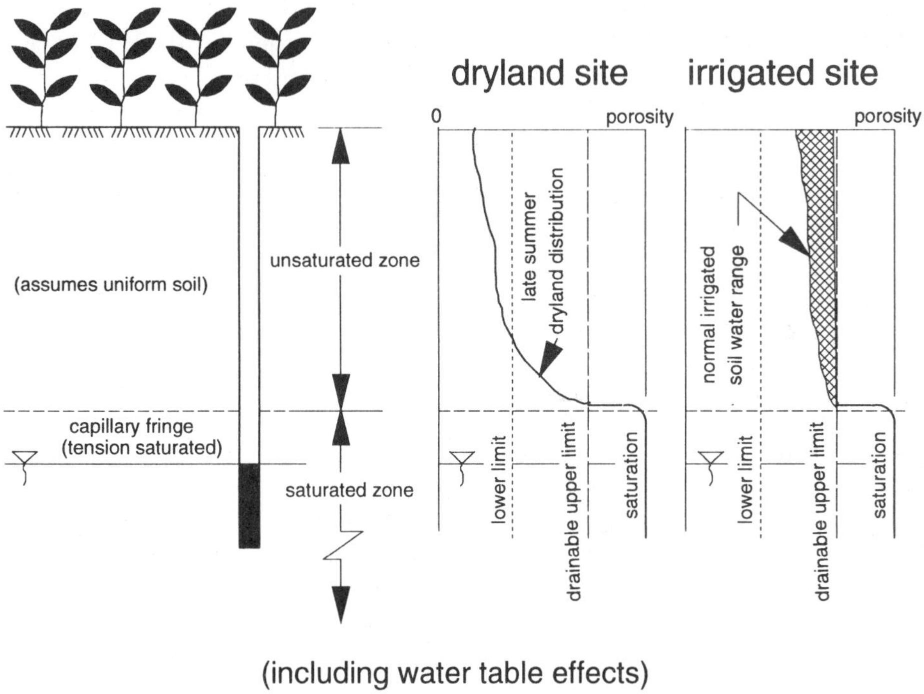

Figure 3.4
Soil water, root zone, and water table conditions

3.5.1 *Darcy's law*

A publication dated in 1856 describes experiments by Henri Darcy to investigate the flow of water through beds of sand for filtration. The results indicated that the flow of water was proportional to the hydraulic head forcing the water and inversely related to the distance traveled. The results have been generalized as follows:

$$v = K\frac{d\psi}{dl} \tag{3.9}$$

where v is defined as *specific discharge* because it represents flow through a space partially occupied by particles and has the units of $m\,s^{-1}$, K is a constant of proportionality and has the same units as v. Both $d\psi$ and dl usually have the units of length, although $d\psi$, as has been discussed often, has a variety of units (Hillel 1980).

Darcy's law has been applied to virtually all porous media fluid mechanics involving water and, in addition, to many other fluids such as natural gas and petroleum products. Water movement in the soil is often multi-phase because of both water and air being present.

3.5.2 *Hydraulic conductivity*

Hydraulic conductivity, K, has the units of velocity and may be determined from both field tests and laboratory determinations. In spite of the fact that conductivity is a term of unusual importance in hydrology, specific instructions for its determination are not presented here.

3.5.3 *Intrinsic permeability*

The hydraulic conductivity, K, is dependent upon the fluid used for its measurement. In an attempt to remove the dependence of K on the fluid, the term *intrinsic permeability* has been introduced.

$$k' = \frac{K\mu}{\rho g} \tag{3.10}$$

where *k'* is known as *intrinsic permeability* and has the units of m^{-2}, μ is absolute viscosity in $Pa \cdot s$, ρ is fluid density in kg m^{-3}, and g is the acceleration of gravity of 9.81 m s^{-2} under standard conditions.

Intrinsic permeability may be used to convert the hydraulic conductivity with respect to one fluid to another; for example, from water to air.

3.5.4 *Saturated soil water movement*

The potential term ψ does not have the term representing the matric potential ψ_m in the case of saturated soil. Water movement in the soil under saturation is much faster than unsaturated soil water movement because the unsaturated conductivity is much slower.

3.5.5 *Unsaturated soil water movement*

The potential term ψ does not have a pressure component in the case of unsaturated flow. The gravity potential is usually small in comparison with matric potential. Matric potential is always negative in sign. Hydraulic conductivity is very small in magnitude and, therefore, unsaturated water movement is usually neglected in most hydrologic calculations of water flux. Nieber and Patwardhan (1989) and Knisel (1980) discuss methods of estimating unsaturated hydraulic conductivity based upon saturated hydraulic conductivity and soil water status. Unsaturated hydraulic conductivity below the DUL (FC) is so small that soil water flow is usually assumed to be zero below the DUL.

3.6 Soil water supply for evapotranspiration

Soil water that supplies evapotranspiration water can come from at least four sources, which are important in making hydrologic analyses. The four sources are natural precipitation, irrigation, stored soil water, and upward water movement from a water table. These are not unrelated because much natural rainfall becomes stored soil water before being lost to the soil profile through ET.

These distinctions are particularly important when discussing irrigation water rights. In the western US an irrigation water right for surface irrigation is really limited to stream-flow depletion caused by irrigation. Of course, irrigation water also comes from underground sources such as wells.

3.6.1 *Stored soil water*

Soil water storage is a function of soil texture with fine textured soils, such as clays, holding more soil water than sandy soils. If winter time recharge of the soil profile is significant, the amount of water supplied to seasonal ET can be significant. Stored soil water has usually been ignored in calculations of irrigation water rights (Jensen et al., 1990).

In the case of commercial crops under irrigation, soil water used by the plants is in the readily available range at all times in an ideal case. In the case of many naturally occurring plants, some ET comes from soil water with water drier than the classic definition of the wilting point intended for commercial crops.

3.6.2 *Natural precipitation*

Much of the water involved with ET can come from natural precipitation and can be either rainfall or snowfall. Natural precipitation often becomes stored soil water before becoming lost to the soil profile through ET. Natural precipitation may of course become part of water in a water table.

3.6.3 *Non-growing season precipitation*

Even in rather severe locations, such as high altitude regions of Colorado and Wyoming, both non-growing season snowfall and rainfall can contribute to stored soil water utilized by plants in the growing season. No real procedures are available for estimating of water amounts stored in this way. Walter et al. (1990) show considerable recharge of soil water within the root zone over the winter for use by non-irrigated plants during the following growing season in a very high dry mountain valley in central Colorado.

Conversely under many conditions soil water may be lost over winter. So called *winter drouth* can be severe enough to kill some plants and is very real indeed.

3.6.4 *Growing season precipitation*

Growing season precipitation that supplies ET is almost always rainfall. The amount used for ET is related to the concept of *effective rainfall*, ER. In the US there are three commonly used models for estimating effective precipitation. A recent reference by Nieber and Patwardhan (1989) recommends evaluating ER by means of a soil water balance model constructed for the purpose. For more information see Section 5.12.1.

3.6.5 *Water supplied by irrigation*

Irrigation is the application of water to a field to prevent, or at least reduce, low soil water effects on plant growth. Nearly all applied water first enters the soil in the root zone before being relocated to the atmosphere by ET. Methods of application involve water application by surface means, sprinklers, or drip systems.

Under the doctrine of prior appropriation followed in the western US, an irrigation water right is limited to the reduction or depletion of stream-flow caused by crop water use or by plants on lands where soil water is increased by the irrigation. In the case of irrigation water originating from underground sources, the limitation is the ET caused by irrigation.

3.6.6 *Movement to or from a water table*

Water movement from a water table to the root zone of plants is unsaturated. This is in spite of the fact that a zone of near saturation called the capillary fringe always exists above a water table. Upward movement of water is limited by unsaturated hydraulic conductivity, which is usually much less than saturated hydraulic conductivity. Some models assure that the hydraulic conductivity is zero below the DUL. This leads to the conclusion that water for ET comes from rainfall, irrigation, or root movement into a moist soil zone.

Chapter 4

ESTIMATING REFERENCE CROP ET

A count of methods for estimating ET has not been made, but the number must be at least 50. Of these, only some of those commonly used to estimate ET_r will be discussed here. Once the more common versions of certain types of equations have been presented, little additional knowledge results from consideration of modifications of the same equations. For example, the Penman equation has many alternative wind functions, but only a few will be presented herein. A common recommendation is to stress the desirability of making local calibrations of methods. Local calibrations will not be discussed here because of space limitations. These alternative calibrations may be very useful and accurate under specific conditions.

Some methods, such as the earlier versions of the Blaney-Criddle method, were intended to predict monthly ET for a specific crop at a specific time. Some methods, such as current versions of the Penman method, have been used to predict reference crop ET, and then a suitable crop coefficient has been used for estimates of vegetative ET. Many citations make the point that it is important to define carefully desired methods used and to keep methodologies consistent (Jensen, 1974; Doorenbos and Pruitt, 1977; Burman et al., 1980,1983; Wright, 1982). This consistency is important for the entire process of predicting ET for specific vegetation. The recommended consistency involves data collection and assembly, the calculation of reference crop ET, and the application of crop coefficients to obtain estimates of crop ET at a specific time and place.

Some methods, such as the Blaney-Criddle and Penman methods, have been through an evolutionary process, and it is important to make sure that the particular version being used is clearly identified along with a careful indication of any required assumptions. This is particularly true for the combination theory methods because of the many thermodynamic, meteorological, and other functions which may be calculated in alternative ways. The Jensen-Haise method received a major modification shortly after its introduction but has remained unchanged since that time. Other methods have seen little change. Words such as *modified* are incomplete without a detailed explanation of specific versions used and assumptions made in applications.

ET values always have the units of depth of water per unit time, mm day^{-1}. Many times evapotranspiration, ET, will be shown as λET which has the units of MJ day^{-1} m^{-2}. The depth unit implies any desired area. At a temperature of 20 °C water has density of 998 kg m^{-3}, For convenience use a water density of 1000 kg m^{-3} (this is only strictly true at 4 °C). ET can be calculated from energy, λET, in MJ m^{-2} day^{-1}, to depth units, mm day^{-1}, by the multiplication by the factor 0.4086. For a detailed analysis of these units see Section 8.2.3.

4.1 Sequence of calculations and flow charts

The subject of Chapter 4 is the prediction of reference crop ET, and it is well to consider the sequence of calculations to be followed in making estimates of *reference crop evapotranspiration*, ET_r. This is for a carefully defined reference crop. There will be some variations in sequence because the hydrologist usually must use historic data which can be presented in a variety of ways. Doorenbos and Pruitt (1977), Burman et al. (1980) and Burman et al. (1983) used flow charts to show sequences of calculations and decisions that need to be followed in order to achieve reliable estimates of ET_r. The following flow charts are constructed to show the sequence of both decisions and calculations required to make reliable estimates of ET_r. The flow charts are intended to show an overview of an usually complex process on a single sheet and to show alternative methods side by side for comparative purposes.

4.2 Potential ET and/or reference crop ET

The term *potential evapotranspiration*, ET_p, has been used for a considerable period. Linsley et al. (1958) describe ET_p as *the evapotranspiration that would occur were there an adequate water supply at all times*. Eagleson (1970) adds qualifications such as *unlimited water to plant stomata*. The term, as such, is very ambiguous since the upper limit of ET is dependent upon the type of vegetation present as well as soil water and climatic conditions. In *Consumptive Use of Water and Irrigation Water Requirements* (Jensen, 1974), the statement is made that ET_p *refers to a well-watered crop like alfalfa*. Jensen et al. (1970) proposed the term *reference crop evapotranspiration*, ET_r replace the term ET_p for alfalfa from 30 to 50 cm in height. In 1977 it was suggested that the reference crop coefficient, ET /ET_{rg}, be applied to grass from 8 to 15 cm. (Doorenbos and Pruitt, 1977).

Under some conditions, the reference crop definition may be difficult to apply. The following discussion is a case in point. Grass grows at high altitudes, but forage grown at high altitudes usually has a high composition of grass-like plants, such as rushes and sedges (Pochop and Burman, 1987). High altitude grass is seldom the same as Alta Fescue grown at Davis, California. Davis data were used in calibration of methods described in FAO-24 (Doorenbos and Pruitt, 1977). The term *grass-reference* by itself is very incomplete. Alfalfa is not commonly grown at elevations of 2000 meters or more.

A specific version of the Penman or Penman-Monteith methods may be selected and used to define reference crop ET for situations where the grass or alfalfa definitions of reference crop ET do not fit. Of course, appropriate crop coefficients would have to be determined by some form of local calibration. Usually such calibration is done under research or intensive study conditions and is not performed under operational conditions.

4.3 Crop coefficients

If the concept of reference crop ET_r is used in estimating ET for a specific crop at a specific time, a suitable crop coefficient must be used. From a practical standpoint crop coefficients are available for the grass definition and the alfalfa definition (Doorenbos and Pruitt (1977); Wright (1982); Jensen el al. (1990). Feddes (1987) presents crop coefficients based on the grass definition using a reference crop ET calculated by the Makkink method. Feddes also published crop coefficients based on potential open water evaporation, E_0. In addition, corrections may be needed to account for a wet soil surface, limited soil water, or other factors which influence crop production and ET. At times these corrections may be incorporated into the crop coefficients.

Alternative definitions of crop coefficients may be made, depending on the method of estimating ET_a.

$$ET_m = K_c ET_r \tag{4.1}$$

$$ET_a = SW\ ET_m \tag{4.2}$$

where ET_a represents *evapotranspiration* from a specific crop at a specific time, including the effects of plant species, growth, and soil water limitations. ET_m is maximum ET when soil water does not limit ET. The *soil water correction*, SW, may include dry soil water conditions and a wet soil surface. K_c is known as *a crop coefficient* and incorporates crop effects. The implementation of these effects is not consistent. Possibly other impacts may be included in K_c. K_c values are determined experimentally, and graphical or other means of computation are specified. The subscript *r* indicates a reference crop. Details of the determination and use of crop coefficients are discussed in Chapter 5.

4.4 Penman methods

Penman (1948) published the first of what has become known as *combination methods*. The method is a result of a combination of a theoretical energy balance with an empirical wind function. The main theoretical limitation is the use of *steady state* equations with a process that obviously involves diurnal or shorter period variations. This assumption is usually not serious for hourly or shorter period calculations. The steady state assumption introduces even more empiricism into its many functions when using daily climatic or longer period average data.

The Penman method has seen considerable evolution throughout the 45 years of its existence. Many assumptions are required for its use. Alternative functions and methods are available. It is, therefore, imperative that a very complete description of all assumptions and all alternative functions be included with the narrative accompanying any use of the Penman method. The modifications for practical applications are continuing at the present time, perhaps at an increasing pace.

The Penman method may be shown as follows:

$$\lambda ET_r = c\left[\frac{\Delta}{\Delta+\gamma}H + \frac{\gamma}{\Delta+\gamma}K_{units}f(u)f(e)\right] \qquad (4.3)$$

$$H = R_n + G \qquad (4.4)$$

where R_n is *net radiation* in MJ day^{-1} m^{-2}, G is *soil heat flux* in MJ day^{-1} m^{-2}, Δ is the slope of the saturated vapor pressure vs temperature relationship in kPa °C^{-1}, γ is the *psychrometer coefficient* in kPa °C^{-1}, *c* is a dimensionless correction factor used originally by Penman (1963) and later by Doorenbos and Pruitt (1977) in the FAO-24 version of the Penman method. The factor *c* is set equal to one in other uses of the Penman method. The factor *c* is usually calculated by a regression developed by Frevert et al. (1983) for the FAO-24 version. The factor *c* varies from near unity to zero as the lower limit. K_{units} has numerical values primarily depending upon desired final units and the input units. The wind function f(u) has the units of m s^{-1} or km day $^{-1}$; and f(e), the vapor pressure deficit term, has the units of kPa. Both will be discussed in detail in following sections. The sign convention used on H assumes that heat transfer is toward the soil surface.

λET has the units of MJ m^{-2} day^{-1}. With the division by the latent heat of vaporization λ the units of ET are now mm day^{-1}. λ has units varying by multiples of k which is 1000, or M which is one million. See Section 8.2.3 for specific details regarding λ and ET. See Section 4.4.13 for details involving units for the FAO-24 version of the Penman method (Doorenbos and Pruitt, 1977).

4.4.1 *Penman correction factor* c

The FAO-24 version of the Penman equation is the only major version in use at the current time where *c* is not set equal to unity (Doorenbos and

Pruitt, 1977). For this version, the coefficient *c* was originally found in a table. Frevert et al. (1983) used multiple regression to develop an empirical equation to estimate values of *c*. Jensen et al. (1990) rounded terms to eliminate the non-contributing decimal. The authors have observed very small differences between the regression equations of ASCE Manual 70 Jensen et al. (1990) and Frevert et al. (1983). The earlier regressions of Frevert et al. are presented because of ease of use and accessibility. Allen and Pruitt (1989) published revised regression equations. The regression equation found in Frevert et al. (1983) follows:

$$c = a_o + a_1 RH_{\max} + a_2 R_s + a_3 U_d + a_4 DN_r + a_5 U_d DN_r + a_6 RH_{\max} R_s U_d + a_7 RH_{\max} R_s DN_r \tag{4.5}$$

$a_o = 0.6817006$
$a_1 = 0.0027864$
$a_2 = 0.0181768$
$a_3 = -0.0682501$
$a_4 = 0.0126514$
$a_5 = 0.0097297$
$a_6 = 0.000043025$
$a_7 = -0.000000092118$

where U_d is daytime *wind velocity* in m s^{-1}, R_s / λ is *solar radiation* in mm day^{-1}, DN_r is the *ratio of daytime to nighttime wind movement*, and RH_{max} is the maximum *relative humidity* in percent.

Usually wind velocity is in km day^{-1}, but the regression equation shown above specifies U_d in m sec^{-1}. The following equation may be used to make the necessary conversion.

$$U_d = \left[\frac{DN_r \cdot u_2}{1 + DN_r}\right]\left[\frac{1000}{12 \cdot 3600}\right] \tag{4.6}$$

where u_2 is in km day^{-1} at an anemometer height of 2 m and U_d is daytime wind velocity in m sec^{-1}.

4.4.2 *Penman wind functions*

Many wind functions f(u) have been developed for use with the Penman combination model to predict reference or potential ET. The wind function is discussed in Section 4.4.2. Discussions of some of the various wind functions are found in Burman et al. (1983) and Jensen et al. (1990). The usual form of the wind function is

$$f(u) = a_w + b_w u_z \tag{4.7}$$

where the constants a_w and b_w are determined by regression during what amounts to a calibration of the Penman method. *Wind velocity, u_z*, is in km day^{-1}, or in some cases, m s^{-1}. A theoretical examination of the wind function reveals that it is related to atmospheric stability, heat and water transfer, and the velocity gradient.

4.4.3 *Seasonally constant wind functions*

The original Penman f(u) was intended to have constant coefficients, a_w and b_w, throughout the growing season. Since the introduction of the Penman method, numerous calibrations have been made and various coefficients have been proposed. Many references, including Jensen et al. (1990), show wind function coefficients in tabular fashion.

4.4.4 *Seasonally variable wind function*

Wright (1982) proposes the following evaluation of Equation 4.7 in which a_w and b_w are not constant but vary throughout the growing season. The following equations are those appearing in Jensen et al. (1990), and the equations are minor variations of those appearing in Wright (1982). These equations are based on u_z being in km day^{-1} at an anemometer height of 2 meters.

$$a_w = 0.4 + 1.4 \exp\left\{-\left[\frac{D-173}{58}\right]^2\right\} \tag{4.8}$$

$$b_w = 0.007 + 0.004 \exp\left\{-\left[\frac{D-243}{80}\right]^2\right\} \tag{4.9}$$

where D is the calendar day number of the day under consideration. For use in the southern hemisphere D should be offset by 6 months.

Consistency of calculations is recommended, and Wright's variable wind function should not be used independently. Wright's variable wind function should only be used with his recommended method of calculating net radiation and other parameters. See the flow chart shown in Figure 4.1. Jensen et al.

(1990) show that the use of Wright's variable wind function and his variable method of estimating net radiation give good results with a worldwide data set.

4.4.5 *Vapor pressure deficit term f(e) or VPD*

The *vapor pressure deficit term*, f(e), abbreviated here as VPD, may be implied as being the following (Penman, 1948):

$$f(e) = VPD = e_T^o - e_e \tag{4.10}$$

where e_T^o is *saturation vapor pressure* in kPa at the ambient *air temperature* T in °C, and e_e is the existing *vapor pressure* in kPa. The time interval was not specified, and the definition is logical and suitable for short term determinations such as periods of an hour or less. However, for data using daily or longer period averages, interpretation of Penman's definition is necessary. The following two sections discuss two different arbitrary interpretations. There are, doubtless, exceptions to the two following approaches.

4.4.6 *VPD, vapor pressure averaging*

Using this definition, the vapor pressure deficit term is based upon the average of the saturation vapor pressures at daily maximum and minimum air temperatures. When applied for longer periods, for example a monthly period, the monthly average daily maximum and monthly average daily minimum air temperatures are used. This approach has been used by Wright (1982) for his calibration of ET_r. The title of this paragraph comes from Cuenca and Nicholson (1982). They indicate that the method of estimating VPD should be consistent with the method of calibrating f(u).

$$VPD_2 = \frac{\left[e_{\max}^o + e_{\min}^o\right]}{2} - e_a \tag{4.11}$$

where $e_{\max}^o$ is the *saturation vapor pressure* at daily *maximum air temperature*, $e_{\min}^o$ is the *saturation vapor pressure* at the daily *minimum air temperature*, and e_a is the daily average *vapor pressure* of the air which is taken as the *saturation vapor pressure* at the average daily *dew point temperature*.

The average daily dew point temperature should represent the mean of several dew point temperature measurements taken periodically throughout a day; for example, the average of 24 hourly values. This procedure is possible using microprocessor controlled weather stations. In many cases, a dew point temperature measurement taken at 8:00 a. m. is a reasonable estimate of the average daily dew point measurement (Jensen et al., 1990).

4.4.7 *VPD, temperature averaging*

Doorenbos and Pruitt(1977) used the following method of calculating the VPD for ET_r. The vapor pressure deficit, VPD_1 is calculated using:

$$VPD_1 = (1 - RH)e^o_{T\,a} \tag{4.12}$$

or

$$VPD_1 = e^o_{T\,a} - e^o_{T\,dp} \tag{4.13}$$

where RH is *relative humidity* expressed as a ratio, the superscript *o* indicates saturation conditions, the subscript *dp* indicates average *dew point temperature* and the subscript *a* is used to indicate daily average air temperature. Vapor pressures are in kPa and temperatures are in °C. See Section 2.2.3 for a discussion of vapor pressure and Section 2.2.5 for a presentation of relative humidity.

4.4.8 *Penman method flow chart*

The flow chart in Figure 4.1 shows three of the many variations of the Penman method. Other variations will have similar computation methods.

4.4.9 *Specific Penman calibrations*

The original Penman method has seen considerable local calibration, and the procedures for determining such parameters as net radiation and soil heat flux have undergone many modifications. This has usually been an evolutionary process.

It is not specific enough to say that the Penman method was used to estimate ET. The narrative describing the use of the Penman method for the estimation of ET should describe the details of procedures used to calculate all parameters.

The various calibrations are too numerous for anything even approaching complete coverage in this volume. Only a few of the most common adaptations are described here. The use of other or more specific adaptations is very similar to those described below. An example of a version of the Penman method calibrated for use in western Nebraska is one described by Kincaid and Heermann (1974).

4.4.10 *1963 Penman (original)*

Penman's original equation required the use of the correction factor, *c*. In 1963 a single stage process where *c* was assumed to be one was proposed. None of Penman's developments specifically recommended the manner of calculating the VPD. It can be assumed that the publications of Penman used the temperature averaging method of determining the VPD. This method is designated as VPD_1, and the method was later adopted by Doorenbos and Pruitt (1977) for reference crop, ET_{rg}. This approach is logical for short term estimates of ET.

The following equation for estimating ET is an adaptation appearing in Jensen et al. (1990). They show two versions of $f(e) = VPD$ although the original Penman paper cited did not utilize the concept of a reference crop, ET_r. The version described in this section may be assumed to estimate ET_{rg} for the reference crop grass (Jensen et al., 1990). They then multiply ET_{rg} by 1.15 to get ET_{ra}.

$$\lambda ET_{rg} = \frac{\Delta}{\Delta+\gamma}(R_n + G) + \frac{\gamma}{\Delta+\gamma}6.43 f(e) f(u) \tag{4.14}$$

$$f(e) = VPD_1 = e^o_{T_a} - e^o_{T_{dp}} \tag{4.15}$$

$$f(u) = 1 + 0.0062 u_2 \tag{4.16}$$

where λET is in MJ m^{-2} d^{-1}, VPD_1 is in kPa, f(u) is dimensionless, and *u* is in km day^{-1}. The form of VPD assumes daily, weekly, or monthly applications.

4.4.11 *Penman equation, Wright and Jensen 1972*

A very extensive set of ET lysimeter measurements from Kimberly, Idaho, lead to the calibration known as the 1972 Kimberly Penman (Wright and Jensen 1972). The $f(e) = VPD_2$ (Section 4.4.6) definition is used. Constant coefficients for the Brunt type equation are recommended using constants, *a*, *b*, a_l *and* b_l, developed for the nearest location to that desired. The f(u) function and the method of estimating R_n follow.

$$\lambda ET_{ra} = \frac{\Delta}{\Delta+\gamma}(R_n + G) + \frac{\gamma}{\Delta+\gamma}6.43 f(e) f(u) \tag{4.17}$$

$$R_n = (1 - \alpha)R_s + R_{nl} \quad (4.18)$$

$$R_{nl} = -[a\ (R_s/R_{so}) + b]\,[a_1 + b_1\sqrt{e_a}]\sigma[(T^4_{k(\max)} + T^4_{k(\min)})/2)] \quad (4.19)$$

$$f(u) = 0.75 + 0.0115u_2 \quad (4.20)$$

where λET, R_n, and G are in MJ m^{-2} day^{-1}, u_2 is in km day^{-1} at an elevation of 2 meters above the ground surface. Vapor pressures, e, are in kPa, and $T_{k(max)}$ and $T_{k(min)}$ are maximum and minimum daily air temperatures in Kelvin for radiation calculations.

4.4.12 *Penman equation, Wright 1982*

Based on the realization that most climatic and plant characteristics change constantly throughout the growing season, Wright (1982) published a calibration of the Penman method with variable wind function $f(u_2)$ constants and a seasonally variable albedo used in estimating net radiation, R_n. Slight modifications of the method appear in Jensen et al. (1990). Even though the calibration was based on conditions prevalent at Kimberly, Idaho, the method performs well under a broad range of climatic conditions (Jensen et al., 1990; Burman, 1985).

The original calibration referred to a shorter growing season, and, therefore, led to periodic functions that produced illogical values of parameters when used for locations having longer or yearly growing seasons. The later variations are designed to overcome this problem.

$$\lambda ET_{ra} = \frac{\Delta}{\Delta+\gamma}(R_n + G) + \frac{\gamma}{\Delta+\gamma}6.43f(e)f(u) \quad (4.21)$$

$$R_n = (1 - \alpha)R_s + R_{nl} \quad (4.22)$$

$$if \quad \frac{R_s}{R_{so}} > 0.7 \quad then$$

$$\alpha = 0.29 + 0.06\sin[30(m + 0.0333\,N + 2.25)] \quad (4.23)$$

$$if \quad \frac{R_s}{R_{so}} \le 0.7 \quad then \quad \alpha = 0.3$$

$$R_{nl} = -[a \cdot R_s/R_{so} + b]\,[a_1 + b_1\sqrt{e_a}]\sigma[(T^4_{k(\max)} + T^4_{k(\min)})/2)] \quad (4.24)$$

$$if\ \frac{R_s}{R_{so}} > 0.7 \quad a = 1.126 \quad b = -0.07 \tag{4.25}$$

$$if\ \frac{R_s}{R_{so}} \leq 0.7 \quad a = 1.017 \quad b = -0.06 \tag{4.26}$$

$$a_1 = 0.26 + 0.1\ \exp\{-[0.0154(30m + N - 207)]^2\} \tag{4.27}$$

$$b_1 = -0.139\ always \tag{4.28}$$

$$f(u) = a_w + b_w u_2 \tag{4.29}$$

$$a_w = 0.4 + 1.4\exp\{-[(D - 173)/58]^2\} \tag{4.30}$$

$$b_w = 0.007 + 0.004\exp\{-[(D - 243)/80]^2\} \tag{4.31}$$

where λET, R_n and G are in MJ m^{-2} d^{-1}; vapor pressures are in kPa, u_2 is in km d^{-1}, D is the *calendar day number* of the year, *m* is the number of the month, and N is the day number in the month under consideration. $T_{k(max)}$ and $T_{k(min)}$ are maximum and minimum daily air temperatures in K.

4.4.13 *FAO-24 Doorenbos and Pruitt, Penman method*

In 1977 the United Nations Food and Agriculture Organization (Doorenbos and Pruitt, 1977) published a calibration of the Penman method which differs from the other calibrations cited here in that an overall correction, *c*, is estimated using climatic parameters which are required for the general usage of the Penman method with the exception of the ratio of day to night wind movement. The correction, *c*, is then used as a self-calibrating coefficient.

Many references make the point that consistency in all features of calculating ET is essential for reliable results. This means that the FAO-24 procedures for calculating net radiation, the VPD, the wind function coefficients, and even the method of calculating crop coefficients, K_c, should all be followed.

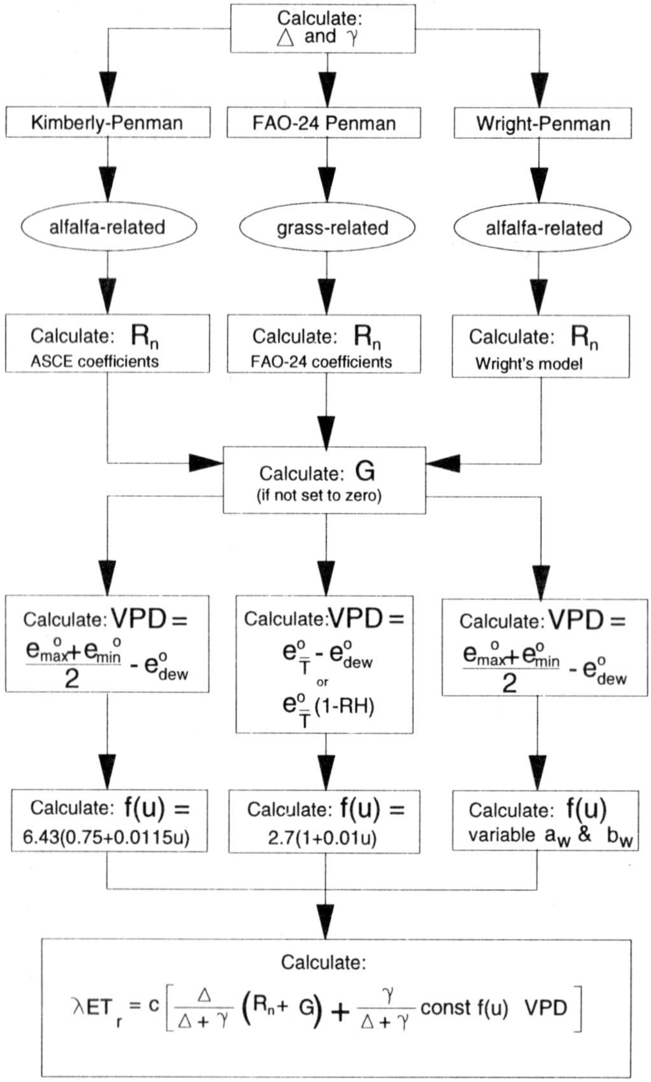

Note: for FAO-24 version calculate: c

otherwise c = 1

Figure 4.1
Penman method flow chart

The FAO-24 definition of ET_r is, with respect to clipped grass from 8 to 15 cm in height, completely shading the ground and never short of water. Data used for calibration came from several locations, but ET and climatic data representing grass alta fescue grown at Davis, California, was obviously of paramount importance. The FAO-24 calibration of the Penman equation follows. In the interest of making equations easy to follow, some redundancy occurs.

$$ET_{rg} = c\left[\frac{\Delta}{\Delta+\gamma}\left\{\frac{H}{\lambda}\right\} + \frac{\gamma}{\Delta+\gamma}2.7f(u)f(e)\right] \quad (4.32)$$

$$H = R_n \qquad G = 0.0 \quad (4.33)$$

$$f(u) = 1 + 0.01u_2 \quad (4.34)$$

$$f(e) = VPD_1 = (e_a^o - e_a) \approx (1 - RH)e_{T_a}^o \quad (4.35)$$

or

$$f(e) = VPD_1 = (e_a^o - e_a) \approx e^o(T_a) - e_{T_{dp}}^o \quad (4.36)$$

$$c = a_o + a_1 RH_{\max} + a_2 R_s + a_3 U_d + a_4 DN_r + a_5 U_d DN_r + a_6 RH_{\max} R_s U_d + a_7 RH_{\max} R_s DN_r \quad (4.37)$$

where ET and R_n/λ are in mm day^{-1} and G is neglected. See Section 8.2.3 for details on conversions between unit systems. Wind velocity u_2 is in km day^{-1}. U_d is day-time wind velocity in m sec^{-1}, DN_r is the day-night wind ratio, RH_{max} is maximum relative humidity in percent, and R_s is solar radiation in the equivalent evaporation unit of mm day^{-1}. U_d may be estimated from wind run in km day-1 and the day night wind ratio using equation 4.6.

The coefficients for estimating *c* follow.

$a_o = 0.6817006$
$a_1 = 0.0027864$
$a_2 = 0.0181768$
$a_3 = -0.0682501$
$a_4 = 0.0126514$
$a_5 = 0.0097297$
$a_6 = 0.000043025$
$a_7 = -0.000000092118$

4.5 Penman-Monteith method

The Penman method was generalized to a significant extent by Monteith (1965). Monteith's variation of the Penman method involves the use of a plant resistance parameter and a more general use of an aerodynamic resistance parameter. Full implementation of the equation is now proceeding at a rapid rate. Recent modifications by Allen et al. (1989) are now being implemented into practical methods of estimating ET. Many additional specific procedures and empirical relationships need to be developed. A specific set of procedures needs to be assembled into a system when a method such as the Penman-Monteith is implemented into practical usage. The original equation was intended for short term calculation such as for an hourly period. Special adaptations are required for daily periods (Allen et al., 1989).

4.5.1 *General Penman-Monteith method*

Monteith (1965) proposed the following addition to the Penman, or more general, combination method of estimating ET.

$$\lambda\,\mathrm{ET} = \frac{\Delta \cdot H + \rho\, C_p\, Q}{\Delta + \gamma^*} \qquad (4.38)$$

$$Q = \left[e^o_{T_z} - e_z\right] / r_a \qquad (4.39)$$

$$H = R_n + G \qquad (4.40)$$

$$\gamma^* = \gamma\left[1 + \frac{r_c}{r_a}\right] \qquad (4.41)$$

where Δ, ρ, γ, and *e* have been discussed in Chapter 2. C_p is the coefficient of *specific heat* for moist ambient air at constant pressure in kJ kg^{-1} K^{-1}, r_a is known as *aerodynamic resistance* in sec m^{-1} to the diffusion of water vapor from the evaporating surface (Thom and Oliver, 1977) and is further defined as the transport resistance between the surface and a height above the surface (Bartholic et al., 1970). The variable *z* is measurement elevation in m, and r_c is canopy resistance in sec m^{-1}.

$$h = \frac{1}{r_a} \tag{4.42}$$

where *h* is known as the *convective heat transport coefficient* discussed in heat transfer literature (Welty et al., 1976) and has the units of W m^{-2} K^{-1}.

4.5.2 *General aerodynamic resistance* r_a

The aerodynamic resistance r_a has many formulations; some are based entirely on empirical evidence, and others are related to the mixing length theory of Prandtl (Daugherty et al., 1977). Most of the expressions utilize a formulation strictly valid for only a stable atmosphere, when actual conditions include both stable and unstable conditions.

Aerodynamic resistance involves friction from air flowing over vegetative surfaces. All fluid friction analysis has a significant empirical component. Satisfactory theoretical methods of fluid friction analyses are unknown to the authors. As a result, all expressions discussed have constants which may be given physical significance but which must be determined experimentally and, therefore, are *fitting constants* to a large extent.

4.5.3 *Basis for aerodynamic resistance equations*

The combination and integration of an expression for turbulent shear and the Prandtl *mixing length* analogy results in the following (Daugherty et al., 1977):

$$u = \frac{1}{k}\sqrt{\frac{\tau_o}{\rho}}(\ln y) + C \tag{4.43}$$

where τ_o is the surface shear in kPa, ρ is *fluid density* in kg m^{-3}, k is *von Karman's constant* which is dimensionless, and *y* is an *elevation* in meters. C is a constant of integration.

Workers in the field of agricultural meteorology used the additional parameters *d* and z_o in evaluating the boundary conditions for the constant of integration when integrating the expression. The parameters *d* and z_o are empirical even though they have physical significance. The additional parameters are needed to describe air velocity profiles over vegetation which may be very complex.

$$u_a = \frac{1}{k}\sqrt{\frac{\tau_o}{\rho}}\ \ln\left\{\frac{z-d}{z_o}\right\} \tag{4.44}$$

where z_0 is defined as the *roughness length* in meters, and d is the *zero plane displacement* in meters. This development assumes an equality between the transfer coefficients of water vapor and momentum.

4.5.4 *Neutral atmospheric conditions*

In 1965 Monteith presented the following expression for r_a for neutral atmospheric conditions. The equation is recommended by Monteith for general use even though it is strictly true only for neutral conditions (Monteith, 1965). A discussion of errors is presented by Monteith. Obviously, general use involves a variety of stability conditions.

$$r_a = \frac{\left[\ln\left\{\frac{z-d}{z_o}\right\}\right]^2}{k^2 u} \tag{4.45}$$

where u is *wind velocity* m s^{-1}, and k is von Karman's constant which is dimensionless. Monteith (1965) did not specify an aerodynamic resistance r_a for unstable atmospheric conditions but discussed methods of adjusting r_a for unstable conditions.

4.5.5 *Unstable atmospheric conditions*

Adding an additional term for unstable atmospheric conditions has been the subject of considerable research. Material in this volume is limited to methods in operational usage. Therefore, the reader is referred to other sources such as Brutsaert (1982) for additional information.

4.5.6 *Other aerodynamic expressions*

Thom and Oliver (1977) recommend the following:

$$r_a = \frac{4.72}{1+0.54u}\left\{\ln\left[\frac{z}{z_o}\right]\right\}^2 \tag{4.46}$$

This expression is said to include the influence of a more realistic surface roughness than Penman's original wind function and even includes calibration incorporating unstable conditions.

4.5.7 *General plant resistance term* r_c

The term r_c was originally defined as r_l, the *leaf resistance*, and then replaced by r_c which has been generalized and interpreted in various ways. A number of subscripts have been used with the term *r*. Allen et al. (1989) define r_c as the *bulk stomatal* or *canopy resistance* s m $^{-1}$; r_c has referred to parts of an individual leaf, the entire plant canopy, and has even included the effects of dry soil (Wallace et al., 1981 and Russell, 1980). The reader should be very careful to verify the interpretation of the term when making ET estimates.

4.5.8 *Penman-Monteith reference or potential ET*

The use of the Penman-Monteith method for estimating reference crop or potential ET requires careful definitions of both the r_a and r_c terms. In addition R_n may require definition. The Penman-Monteith method has been used for the direct estimation of ET for a specific vegetative surface at a specific time and place. Direct estimation of ET does not involve the intermediate step of first estimating reference crop, ET_r, and then the application of a suitable crop coefficient to estimate ET. The use of the Penman-Monteith method in the direct estimation of ET is discussed in Chapter 5.

Specific aerodynamic resistance $\mathbf{r_a}$

Jensen et al. (1990), Allen et al. (1989) and Allen (1986) recommend the following expression for r_a:

$$r_a = \ln\{z_w - d/z_{om}\} \ln\{z_h - d/z_{oh}\}/k^2 u_z \quad (4.47)$$

where r_a is aerodynamic resistance in sec m^{-1}, z_w is the anemometer height, and z_h is the height of the humidity measurement both in meters. The parameter *d* is known as the *roughness length* in meters, and z_{om} is the *roughness length* for momentum or more specifically for wind, and z_{ov} is the *roughness length* for vapor and heat transfer or humidity both in meters.

The parameter *d* is often assumed to be zero; however, Allen et al. (1989) recommend the following empirical relationship for the reference crop alfalfa.

$$z_{om} = \frac{z_v}{8.15} = 0.123 z_v \quad (4.48)$$

$$z_{ov} = 0.1 z_{om} \tag{4.49}$$

$$d = \frac{2}{3} z_v = 0.67 z_v \tag{4.50}$$

where z_v is the forage *height* in meters.

***Specific plant resistance* $\mathbf{r_c}$**

The Irrigation Water Requirements Committee of ASCE (Jensen et al., 1990), Allen et al. (1989) and Allen (1986) recommend the following for r_c:

$$r_c = \frac{r_l}{0.5 LAI} \tag{4.51}$$

where the *leaf area index*, LAI, is the area of plant leaves divided by the area of ground surface in question; r_l is the average resistance of an individual leaf and has a value of about 100 s m^{-1} under the moist soil water status recommended for reference crops. LAI (Allen et al. 1989) may be estimated by:

$$LAI = 5.5 + 1.5 \ln(z_v) \tag{4.52}$$

$$LAI = 24\, z_v \qquad \textit{clipped grass} \tag{4.53}$$

where z_v is *forage height* in meters.

For estimating ET_r Allen et al. (1990) and Jensen et al.(1990) assume that $e^o_{T_z}$ is the average of the saturation vapor pressure at the maximum and minimum air temperatures in kPa. See Section 8.4.7 for an example of calculations.

4.5.9 *Direct estimates of ET using the Penman-Monteith model*

Direct estimation of ET_{max} using the Penman-Monteith model is described in detail in Chapter 5 *Estimating ET for specific crops*. The method uses specific r_a, r_c, R_n, and G values for a given plant at a particular time. Then a soil water model must be used to make a final estimate of ET_a. Thompson et al. (1981) describe a comprehensive system used in the British Islands. See Chapter 5 for more details.

4.5.10 *Penman-Monteith method flow charts*

Figure 4.2 shows a flow chart comparing both the crop coefficient version of the Penman-Monteith method (Allen et al., 1989) and the direct calculation version (Thompson et al., 1981).

4.6 Radiation methods

Recognition that the ultimate source of energy required for ET is the sun lead to the origination of the Makkink method (Makkink, 1957) and to the Jensen-Haise method (Jensen and Haise, 1963). Doorenbos and Pruitt (1977) re-calibrated the Makkink method in FAO-24.

4.6.1 *Jensen-Haise method*

The Jensen-Haise method, commonly used in the western US, was first published in 1963 (Jensen and Haise). The Jensen-Haise method pre-dates the widespread use of true crop coefficients, but the equations shown below have been used with alfalfa-related crop coefficients for some time (Jensen, 1974). Jensen-Haise equation coefficients for the estimation of grass evapotranspiration, ET_{rg}, are available (Jensen and Haise, 1963) but are seldom used.

$$\lambda\, ET_{ra} = C_T(T - T_x)R_s \tag{4.54}$$

$$C_T = \frac{1}{(C_1 + 7.3C_H)} \tag{4.55}$$

$$C_H = \frac{5.0\text{ kPa}}{(e_2 - e_1)} \tag{4.56}$$

$$C_1 = 38 - \frac{z}{152.5} \tag{4.57}$$

$$T_x = -2.5 - 1.4\,(e_2 - e_1) - \frac{z}{550} \tag{4.58}$$

where T is the average *air temperature* for the period °C, e_2 and e_1 are *saturation vapor pressures* at the average monthly *maximum air temperature* and monthly *minimum air temperatures*, respectively, for the warmest month of the year, kPa, and z is the *elevation* above sea level in meters. The evaluation of e_1 and e_2 should be based upon long term mean climatological data.

The Jensen-Haise method has been used for periods ranging from 1 day to monthly. ASCE recommends its use for a minimum time period of 5 days (Jensen et al., 1990).

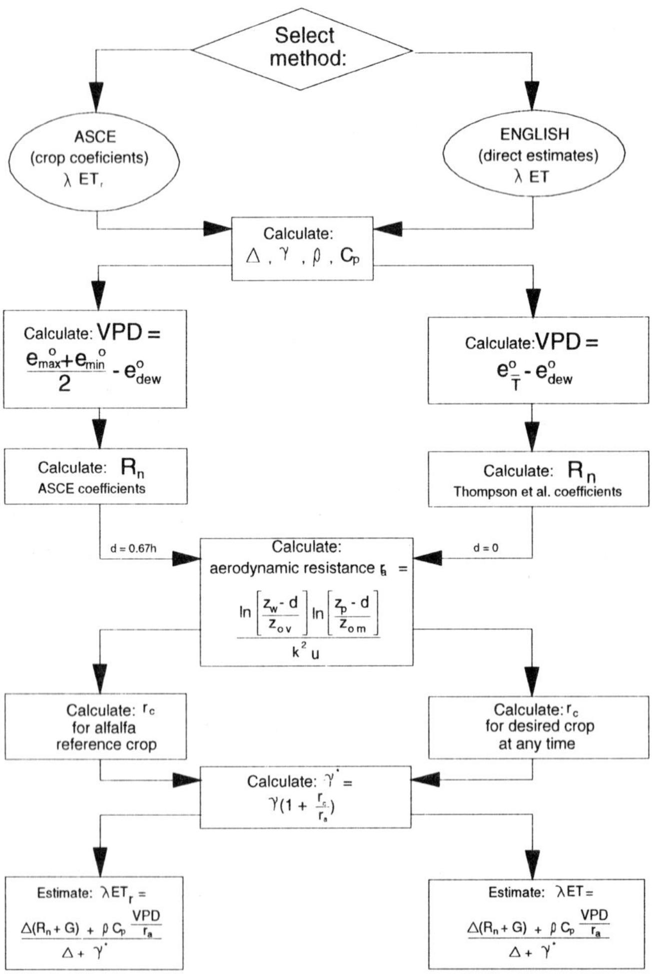

Figure 4.2
Penman-Monteith method flow chart.

4.6.2 *Makkink and FAO-24 radiation methods*

Makkink (1957) originated a method of estimating ET, using R_s for a short green crop shading the ground and never short of water.

$$ET = 0.61\left\{\frac{\Delta}{\Delta+\gamma}\right\}\frac{R_s}{58.5} - 0.12 \tag{4.59}$$

where R_s is in equivalent energy units, ly day^{-1}, and ET is in mm day^{-1}.

In a new calibration, Doorenbos and Pruitt (1977) show an adaptation of the Makkink method as:

$$ET_{rg} = a + b\left[\frac{\Delta}{\Delta+\gamma}\right]\frac{R_s}{\lambda} \tag{4.60}$$

where *a* and *b* are empirical factors which account for mean *relative humidity* and daytime *wind movement*. ET and R_s/λ are in mm day^{-1}.

Frevert et al. (1983) developed a series of relationships using multiple correlation techniques for the determination of *b*. Their equation is very useful for ET estimates using a computer for calculations. Later regression equations are presented by ASCE (Jensen et al., 1990). Because the authors have found little difference in results and because of simplicity and accessibility, the equations of Frevert et al. (1983) follow:

$$a = -0.3 \quad \text{for all cases} \tag{4.61}$$

$$b = a_o + a_1 \cdot \overline{RH} + a_2 \cdot U_d + a_3 \cdot \overline{RH} \cdot U_d + a_4 \cdot \overline{RH}^2 + a_5 \cdot U_d^2 \tag{4.62}$$

a_o = 1.0656
a_1 = -0.0012795
a_2 = 0.044953
a_3 = -0.0002033
a_4 = -0.000031508
a_5 = -0.0011026

where $\overline{RH}$ is mean *relative humidity* in percent, and U_d is *daytime wind velocity* m s^{-1}.

4.7 Temperature methods

A practicing hydrologist is faced with *making do* with available climatic data. Historically, in the US, most climatic data has been collected by a network of cooperative volunteer observers. Only a few weather stations, such as those at airports, collected data sufficient for combination methods. For most stations in the US, daily maximum and minimum air temperatures and daily total precipitation have been, and still are, the only climatic data collected. For this reason, a method of estimating ET which requires only air temperature data proves to be very useful, particularly for historic data. The increasing use of microprocessor controlled climatic measurement systems is changing climatic data collection.

Recently, agricultural climatological networks have collected the required data for combination equations. Calculations are possible for time periods of from one day to less than one hour. Examples are the extensive network of microprocessor controlled climatological stations originally known as *AGNET* that have been operated in Nebraska and surrounding states, and the British network known as *MORECS*. Data from these networks, as well as computerized computations, have made sophisticated methods of estimating ET much more practical for operational usage.

Several temperature methods of estimating ET have been developed. These include the Blaney-Criddle methods which follow, and methods proposed by Hargreaves, Christiansen, Thornthwaite, and others. The Thornthwaite method (1948) is only useful as a climatic index and not for operational estimates of ET. Hargreaves methods still require local calibration (Hill et al. 1989). The Blaney-Criddle method represents established practice over many years and with local calibration (Jensen et al. 1990 and Hill et al. 1989) can provide very useful estimates of long term average ET.

4.7.1 *Original Blaney-Criddle method*

The Blaney-Criddle method was introduced in 1942 and became popular in the 1950's. Because of its simplicity and the requirement of only air temperature measurements, the Blaney-Criddle became very popular in practical situations and remains widely used. The original Blaney-Criddle method is defined as (Blaney-Criddle, 1962):

$$U = K_c F \tag{4.63}$$

where U is seasonal water use in inches. Historically, K_c is an *empirical seasonal coefficient*, and F is the sum of the monthly factors *f*.

$$f = \frac{Tp}{100} \tag{4.64}$$

where T is average *air temperature* in °F, and p is the *monthly percentage of daytime in the year;* p may be estimated using day length models described in Chapter 2. See Chapter 2 for tabular values and algorithms suitable for the determination of the percent daytime. The accuracy of estimates of p is discussed in Chapter 8.

The *empirical seasonal coefficient* is not being used at the present time. Blaney and Criddle (1962) further define the method for monthly determinations of ET as follows:

$$u = k\frac{Tp}{100} \tag{4.65}$$

where u is now monthly ET in inches, and T is average air temperature in °F. The symbol k is designated as the empirical *crop growth stage coefficient.* Only the relatively recent FAO version of the Blaney-Criddle method utilizes the two stage process involving the calculation of reference crop ET, and, for this reason, earlier versions of the Blaney-Criddle method are discussed in detail in Chapter 5. Methods for estimating day length as a function of latitude and date are discussed in Chapter 2.

In a general, perhaps a crude, manner the $u = Tp/100$ may be viewed as a form of ET_p or ET_r with the k values then being viewed as a form of a crop coefficient. It should be reinforced that the Blaney-Criddle k values include a climatic effect (Jensen et al., 1990). The Blaney-Criddle method is still being widely used for legal deliberations in the US. Again, careful definition of methods used is essential.

4.7.2 *SCS Blaney-Criddle method*

The US Soil Conservation Service made a significant revision to the Blaney-Criddle method (USDA, 1970). The reader is referred to Chapter 5 for a complete discussion of the SCS version of the Blaney-Criddle method.

4.7.3 *FAO-24 Blaney-Criddle method*

The FAO-24 version of the Blaney-Criddle method represents a major modification because climatic information, in addition to the required air temperature data, is included. Climatic information includes minimum humidity, sunshine, and daytime wind movement. The additional climatic information may be based on subjective knowledge rather than measured information. The climatic classifications may be based upon experience gained from living and working in an area. If the climatic information comes from actual measurements, a combination method such as the Penman equation is usually regarded as being superior.

All previous versions of the Blaney-Criddle method did not involve the use of a reference crop with the calculation of ET_r in a formal carefully defined manner. In FAO-24 (Doorenbos and Pruitt, 1977) the use of ET_r with crop coefficients was introduced into the Blaney-Criddle method. Thus, in principle, crop coefficients for the grass-based method of estimating ET_r can be used in a two-step calculation of ET for specific vegetation at a specific time, provided the crop coefficients are grass-based and are comparable with applications like FAO-24. Details of the FAO-24 version of the Blaney-Criddle method are described by Doorenbos and Pruitt (1977).

$$ET_{rg} = a + bf \tag{4.66}$$

$$f = p(0.46T + 8.13) \tag{4.67}$$

$$a = 0.0043RH_{min} - n/N - 1.41 \tag{4.68}$$

$$b = a_o + a_1 RH_{min} + a_2 n/N + a_3 U_d + a_4 RH_{min} n/N + a_5 RH_{min} U_d \tag{4.69}$$

where E_{rg} is reference crop ET in mm day^{-1}, ET_{rg} is defined as being equivalent to ET from grass from 8 to 15 cm. in height, p is the *percent of daytime hours* for a single day compared to the day-length of an entire year, and T is average air temperature in °C. The quality n/N is measured sunshine divided by possible sunshine and is dimensionless. RH_{min} is minimum relative daily humidity in percent. U_d is daytime wind speed in m s^{-1}. The coefficients in the regression equation are recommended by Frevert et al. (1983) as:

a_o=0.81917
a_1=-0.0040922
a_2=1.0705
a_3=0.065649
a_4=-0.0059684
a_5=-0.0005967

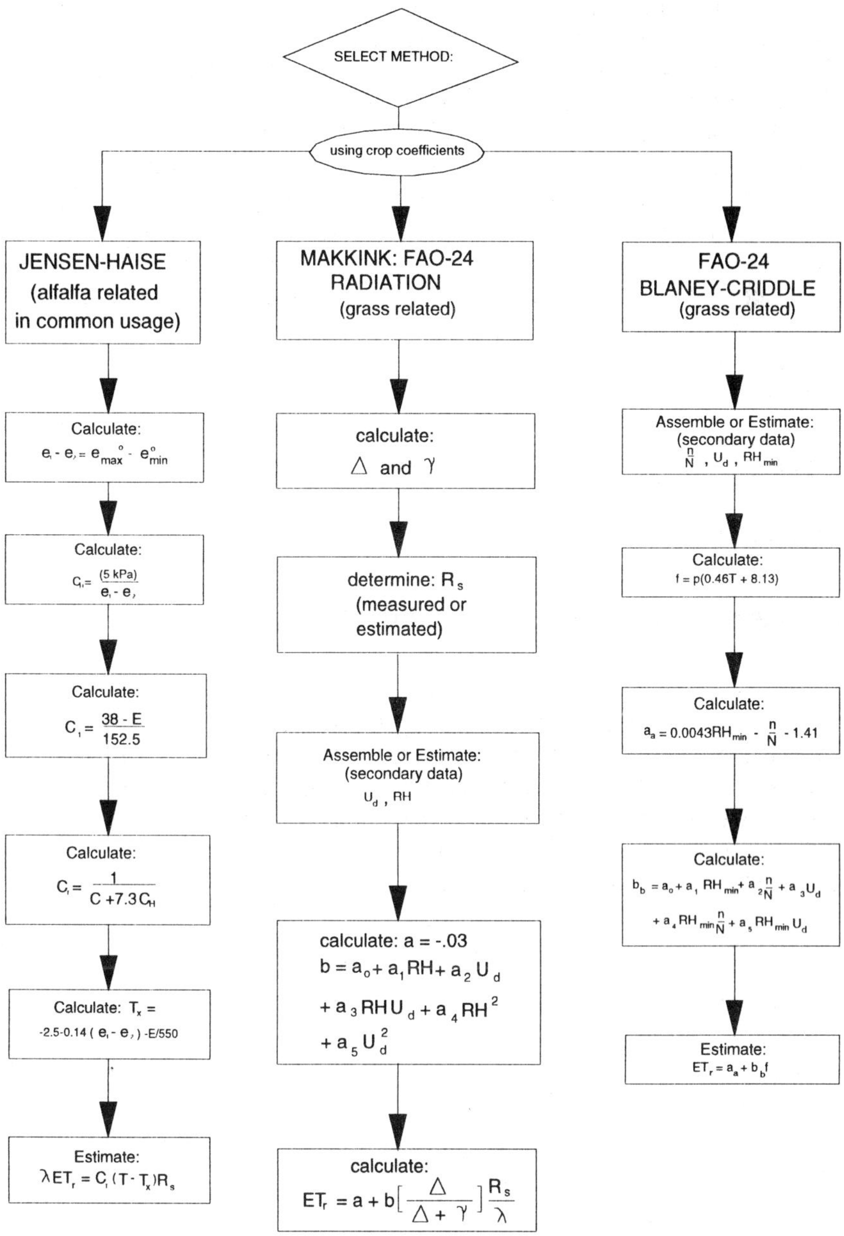

Figure 4.3
Flow chart for Jensen-Haise, Makkink and FAO-24 Blaney-Criddle methods.

4.8 Radiation and temperature methods flow charts

Figure 4.3 is a flow chart illustrating the sequence of calculations that are recommended for radiation and temperature methods used both with crop coefficients.

4.9 Estimating reference ET using measured pan evaporation

Evaporation from an open container such as a Class A evaporation pan provides a simple method of estimating ET. Several cautions are needed, though.

4.9.1 *Pan evaporation as an index of ET*

Water loss from an open container such as an evaporation pan and ET from vegetation both involve the process of evaporation. It is almost always easier to collect data from an evaporation pan than from vegetation.

Virtually all ET studies are made with supporting climatic data being collected. Pan evaporation is a very common measurement and provides excellent supporting data useful for correlation and prediction. In addition, pan evaporation is commonly measured at reservoir and lake sites to be used in estimating water surface evaporation losses.

Because both pan evaporation and ET involve the same basic process, it is easy to assume that a reasonable estimate of reference ET might be found by multiplying measured pan evaporation by a factor usually less than unity. The general relationship is

$$\mathrm{ET}_r = K_p E_p \tag{4.70}$$

where E_p is measured *pan evaporation* in any desired units, and K_p is known as a *pan coefficient* which is dimensionless and generally varies from zero to near unity.

Since pan data is being used to estimate reference crop evapotranspiration, the magnitude of the pan coefficient will be dependent upon whether grass or alfalfa reference ET is being sought. Doorenbos and Pruitt (1977) gave pan coefficients to estimate grass-reference. Their coefficients are for Class A pan data and consider different ground covers, level of mean relative humidities, and 24-hour wind runs. The coefficients of Doorenbos and Pruitt (1977) appear in Table 4.1.

Table 4.1
Suggested values of K_p for Class A pans
ET_r for grass 8-15 cm tall (Doorenbos & Pruitt, 1977)

		CASE A			CASE B		
		SURROUNDED BY SHORT GREEN CROP			SURROUNDED BY DRY BARE AREA		
	fetch dist	mean relative humidity %			mean relative humidity %		
wind conditions *	(m)	< 40	40-70	> 70	< 40	40-70	> 70
light	1	0.55	0.65	0.75	0.70	0.80	0.85
	10	0.65	0.75	0.85	0.60	0.70	0.80
	100	0.70	0.80	0.85	0.55	0.65	0.75
	1,000	0.75	0.85	0.85	0.50	0.60	0.70
moderate	1	0.50	0.60	0.65	0.65	0.75	0.80
	10	0.60	0.70	0.75	0.55	0.65	0.70
	100	0.65	0.75	0.80	0.50	0.60	0.65
	1,000	0.70	0.80	0.80	0.45	0.55	0.60
strong	1	0.45	0.50	0.60	0.60	0.65	0.70
	10	0.55	0.60	0.65	0.50	0.55	0.65
	100	0.60	0.65	0.70	0.45	0.50	0.60
	1,000	0.65	0.70	0.75	0.40	0.45	0.55
very strong	1	0.40	0.45	0.50	0.50	0.60	0.65
	10	0.45	0.55	0.60	0.45	0.50	0.55
	100	0.50	0.60	0.65	0.40	0.45	0.50
	1,000	0.55	0.60	0.65	0.35	0.40	0.45

* light: < 175 km/day, < 2 m/s
moderate: 175-425 km/day, 2-5 m/s
strong: 425-700 km/day, 5-8 m/s
very strong: > 700 km/day, > 8 m/s

Pan coefficients for estimating alfalfa reference ET are not widely available. Jensen et al. (1990) provide monthly ratios to convert other references, including pan references, to alfalfa reference ET. However, they caution that these ratios are calibrated only for Kimberly, Idaho, conditions, and extreme caution should be exercised in considering their use at other locations.

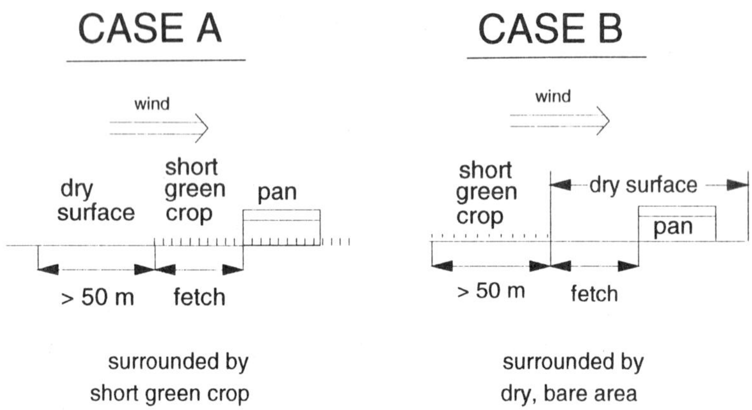

Figure 4.4
Illustration of evaporation pan environment
from Jensen et al. (1990)

4.9.2 *Evaporation pans compared to vegetation and open water*

E from evaporation pans is said to be more responsive to surrounding conditions than ET from fields containing vegetation. Vegetation has many important differences with respect to evaporation pans.

1. Albedo, α, of green plant surfaces is from 0.2 to 0.25, while water surface albedo varies from 0.02 to a maximum near 1.0, depending on the angle of incidence of solar radiation and many other factors. The other factors include the depth of the water, type of clouds, and wind velocity . The α of plants may not to be as sensitive to the angle of incidence of solar radiation as an open water surface.

2. Heat storage in soil differs from evaporation pans.

3. Roughness to air movement differs.

4. Exposure to climate differs widely.

5. Pan evaporation lags ET on a diurnal basis.

6. More pan evaporation may occur at night compared to ET.

Bodies of water of practical size for the estimation of E are massive compared to any evaporation pan used in a climatological study. Several important differences emerge.

1. Temperature variations with depth differ.

2. Storage of heat within the pan can be appreciable and may differ from reservoirs, lakes or ponds.

3. Exposure to climate.
   a. Fetch should differ.
   b. Larger bodies of water will modify air flowing over them.

4. Differences in turbulence, temperature, and humidity of the air immediately above the water surface versus the evaporation pan surface can make a difference.

5. Heat transfer through the sides and bottom of the pan occurs. This is different from the heat transfer from a reservoir, lake, or pond.

6. The size of pan, differences in the surroundings, use of cover screens, and different pan colors affect water loss. Some of these effects, for example the size and surroundings, may be similar to the effects on water losses from reservoirs, ponds, or lakes.

4.9.3 *Types of pans*

Many different types of evaporation pans have been proposed and used all over the world. Principal variations are

1. Size (diameter, depth).

2. Placement (on surface or buried).

3. Covering (screened or non-screened, etc).

4. Construction materials.

The size variations are large, varying from a Class A evaporation pan of 1.219 m (4 feet) in diameter to the use of a common laundry wash pan used for irrigation scheduling in Montana. The *wash pans* were proposed as being a more economical alternative to relatively more expensive specially manufactured pans.

Buried and screened pans have been proposed as providing evaporation measurements that are more similar to ET or E from non-vegetative water surfaces. Variations in construction materials also have been proposed to make measurements closer to ET, or water body evaporation.

Most of the measurements made in the US utilize the US Class A evaporation pan, and it is in common usage in many other parts of the world. Hounam (1973) presents factors for converting pan E from one type of pan to another. Such conversions should be viewed as approximations only. Burman (1976) discusses methods of estimating pan E for the purpose of estimating missing data.

The National Weather Service of the US (US Department of Commerce, 1982 a,b) published a summary and atlas of Class A pan evaporation for the US, using both measurements and climatological estimates. They show that pan evaporation decreases with elevation above sea level.

4.9.4 *Operation of evaporation pans*

It is rational to assume that data from an evaporation pan not operated under standard conditions will be questionable. Detailed instructions for the operation of the US Class A pan may be found in recommendations by the National Weather Service (US Department of Commerce, 1972) and later in the National Handbook of Recommended Methods for Water-Data Acquisition (US Department of the Interior, 1981).

4.9.5 *Limitations of evaporation pan data*

Lewis (1978) reports on the use of data from Class A pan sites in Wyoming. He indicates that some data from sites were not usable because of poor maintenance and/or poor exposure. In addition to pan evaporation sites, the authors recall many instances of similar operational errors or changes in exposure in climatic station sites. New buildings are constructed, and streets or roads are placed in such a manner as to influence climatic data. Such changes or errors are certainly a hazard for evaporation pan stations.

When installing evaporation pans, it is important to follow recommendations. When using historical data, it is important to try to verify that installation conditions were according to standards and that operational procedures followed recommendations.Evaporation pans are included as part of some weather stations, especially important agricultural weather stations. Evaporation pans provide a measurement of the integrated effects of temperature, radiation, wind, and humidity on evaporation from a specific open water surface. In a similar fashion, plants respond to the same climatic variables, but several major factors may produce significant differences in loss of water. Evaporation pans should be located in agricultural areas with the immediate surroundings preferably of a green and frequently mowed grass cover. Dry bare soil or concrete surroundings can increase water loss from pans, whereas covered screens are expected to reduce water loss. The type of pan is also very important. In the US, most pans are Class A pans. These are 1.219 m in diameter by 0.3048 m deep made of galvanized iron, stainless steel, or Monel metal mounted level on a wooden platform about 0.2169 m above the ground surface. The water level should be maintained 51 mm to 76 mm below the rim.

4.9.6 *Christiansen method using pan evaporation*

Christiansen (1968) and Christiansen and Hargreaves (1969) presented an equation for estimating potential ET from Class A pan and climatic data. Since data from grass were used to calibrate their equation, the approach might be considered as estimating grass-reference ET. Their equation is:

$$\mathrm{ET}_r = 0.755 E_{pan} C_T C_W C_H C_S \tag{4.71}$$

where E_{pan} is measured *pan evaporation*, and the C terms are dimensionless coefficients, given as:

$$C_T = 0.862 + 0.179\left[\frac{T}{T_o}\right] - 0.041\left[\frac{T}{T_o}\right]^2 \tag{4.72}$$

$$C_W = 1.189 - 0.240\left[\frac{W}{W_o}\right] + 0.051\left[\frac{W}{W_o}\right]^2 \tag{4.73}$$

$$C_H = 0.499 + 0.620\left[\frac{H}{H_o}\right] - 0.119\left[\frac{H}{H_o}\right]^2 \tag{4.74}$$

$$C_S = 0.904 + 0.008\left[\frac{S}{S_o}\right] + 0.088\left[\frac{S}{S_o}\right]^2 \tag{4.75}$$

where T is the mean *air temperature*, and $T_0 = 20°$, both in °C; W is the mean *wind velocity* 2 meters above the ground, and $W_0 = 6.7$, both in km per hour; H is the mean *relative humidity*, and $H_0 = 0.60$, both expressed as decimals; and S is the *percentage of possible sunshine*, and $S_0 = 0.80$, both expressed as decimals.

Chapter 5

ESTIMATING ET FOR SPECIFIC CROPS

The final goal of predicting ET is to obtain ET estimates for specific crops at a given time or over a defined time period. Estimating ET for a specific crop can be a very complex matter, depending upon the amount of refinement desired. Some of the factors affecting crop ET include the type of crop, stage of growth, climate, soil water, fertilizer level, soil type and condition, weed competition, plant population or cover, and pests and diseases. From a practical standpoint, it is generally impossible to consider directly all of these factors when estimating crop ET. Usually ET estimates consider the climate, plant, and soil water effects. Omission of the other factors accounts for much of the spatial and temporal variations that may occur in ET estimates, even when using the same methodology.

The effects of crop characteristics on ET of a specific crop are very important and generally are accounted for by the use of empirically derived crop coefficients. Chapter 4 describes the calculation of reference crop ET, which is usually designated as ET_r and is for a hypothetical crop defined as well-watered alfalfa or grass, usually with cutting effects removed and vegetation height limitations. Methods for calculating reference crop ET consider the climatological effects. Evapotranspiration for a well-watered crop, other than alfalfa or grass, is estimated by multiplying the reference crop ET by a crop coefficient, K_c, which is the ratio of the water use of the crop to that of the reference crop. Occasionally, pan evaporation is used as an index for estimating crop ET or reference crop ET.

Some methods of estimating ET do not employ carefully defined crop coefficients. For example, most versions of the Blaney-Criddle method use what are called crop growth stage coefficients, which are not the same as the crop coefficients applied to grass or alfalfa reference ET. See the following section for more discussion on crop coefficients. Thompson et al. (1981) describe the use of direct calculations of crop ET for a network of stations in the United Kingdom, using the Monteith version of the Penman equation.

Soil water becomes an especially important factor when levels are low enough to limit crop growth. Estimation of crop ET for limited soil water conditions is even more complex than for well-watered crops. Generally, ET estimates for well-watered crops are adjusted in some fashion to account for limited soil water conditions.

Even in the most complex method of estimating ET_r, some climatic factors either are not a part of the theory used or are not measured. The influences of these unmeasured factors become absorbed in the empirical relationships used in all ET estimation procedures. A method of estimating ET should not be used under conditions that are significantly different from conditions prevalent at sites whose data were used for calibration of the method.

It is also axiomatic that any change in crop growth will result in a change in ET. The changes in ET may be major or modest in magnitude. The following is a discussion of crop environmental considerations which may influence ET.

5.1 ET using ET_r

The reference crop used in methods for estimating crop ET must be designated for each method, and coefficients for that particular reference crop must then be used. Grass and alfalfa are most commonly used as reference crops. However, as compared to methods of estimating reference crop ET, much less work has been done in developing crop coefficients and, thus, the uncertainties and variations in crop coefficients may be expected to be relatively large. For example, Robb et al. (1984) state: *Much work has been published on the reference calculation (alfalfa or grass water use) or measurement (pan evaporation), while crop curve data are not widely available.* Also Rijtema and Aboukhaled (1975) state: *The derived crop coefficients used in these formulae are changing under different growing conditions, which must lead to the conclusion that they are only valid under the conditions of the performed experiments.*

Another difficulty in applying many empirical methods of estimating crop ET is that not all methods have defined clearly the reference for which they have been derived. The original Blaney-Criddle method, for example, although not considered by most to be a method which applies crop coefficients, does provide an estimate of reference ET. This estimate of reference ET is obviously imperfect but is still an estimate of reference ET. Reference ET is then multiplied by the proper crop growth stage coefficients for each particular crop to obtain ET_a. Thus, the methodology for estimating crop ET is the same as those methods using either grass or alfalfa as a reference. The difference is that coefficients applicable specifically to the Blaney-Criddle must be used (e.g., Jensen, 1974).

5.2 Botanic and cultural influences

The observant person soon notices that crops respond to the combined soil, chemical, insect, and other environmental considerations where they grow. Siemer and Rumberg (1975) report that, in spite of the crop species planted in meadows in the high altitude areas of the Rocky Mountains of the US, the vegetative composition changes to those species that are suited to the water and fertility conditions found at the site.

Of the following factors, only soil water is directly utilized in most methods of estimating ET. Therefore, the reader is cautioned to remember that unmeasured factors are present. The question of suitability is always relevant. Some computerized plant growth models, however, include more of the considerations.

5.2.1 *Growth cycles*

The normal growth cycle of a plant has a significant influence on ET. The effects of the growth cycle are usually incorporated in typical crop curves used to estimate ET for a specific crop at a specific time.

The separation of the growing season into intervals is obviously an arbitrary process. The subdivisions are usually tailored to the kind of crops and the method of estimating ET. The growth stage subdivisions for a field crop like corn are obviously different than those of riparian vegetation such as willows.

The detailed subdivisions of Wright (1982) have been generally followed in this volume. The categories are arbitrary and, therefore, will vary in other references. In the case of direct calculation of ET for a specific time and place, the various coefficients used must reflect the influences of growth stage on ET. The discussion in this section is not intended to be comprehensive but appears for the purpose of emphasizing that growth stage is important in both vegetative management and ET prediction. Specific growth stages are discussed more fully in the following paragraphs which deal with crop curves used in the prediction of specific ET. Growth stage discussions follow.

Planting and emergence

Crops are placed in or on the soil mechanically in the case of commercial crops or by natural means in the case of uncultivated vegetation. The term *emergence* is self-defining. At that time the soil surface may be bare. ET will be primarily *evaporation*, E, from the soil surface with *transpiration*, T, from plants being zero or small in magnitude.

Special water management techniques are utilized in the case of commercial crops. Often irrigation water is supplied so that the crops will have sufficient water for germination and emergence of the crops.

Rapid growth

This growth stage is primarily characterized by an increase in vegetation which is defined by the term *leaf area index*, LAI. There is little change in color.

Full cover

Full cover is usually when an observer above the field would only see vegetation. Sometimes full cover is a few days later than the start of the heading or blooming growth stage (Wright, 1982).

Heading or bloom

In many cases the crops in this growth stage are regarded as being the most sensitive to drouth. In an early classic, Rhoades and Nelson (1955) refer to this growth stage as the *tasseling and silking stage* for corn.

Ripening and harvest

Crops often undergo a significant change in the color of vegetation which is accompanied by leaves drying and even changing structure. Soil water control is important because of the need for trafficability during harvest operations. Drouth in this stage usually has only minimal influence on crop yield, but soil water level may be important because of its relationship to quality, sugar content, or other similar quality.

Perennial and biannual crops

Perennial crops undergo the establishment stage only once during their lifetimes. Many trees, grasses, and other crops are examples of naturally occurring crops that are usually perennial in character. Alfalfa is a primary example of a perennial commercial crop. Biennial crops undergo a combination of the annual and perennial growth stages.

Determination of growth stage dates

Wright (1982) and ASCE Manual 70 (Jensen et al., 1990) show tables of dates of occurrence of various portions of growth stages for Kimberly, Idaho. Doorenbos and Pruitt (1977) have a table showing suggested lengths of their designations for growing season. The US Soil Conservation Service has produced *irrigation guides* for various states or areas which include estimates of dates of occurrence of different parts of the growing season. It may be assumed similar summaries exist for other parts of the world.

For the operational prediction of growing season dates, the user is encouraged to gather all pertinent information from governmental agencies. In addition, suggestions should be sought from any research workers who are familiar with the area. The experience of farm operators in the area is also invaluable. All relevant information should then be summarized, and final decisions should be based on all information and common sense.

5.2.2 *Reference or potential ET*

Reference or potential ET is discussed in several places in this volume and therefore, duplication has occurred. The reader is referred specifically to the following sections for alternative discussion, Sections 1.3.3,4.2, and 5.1.

Often the words *potential* or *reference* are used in connection with the estimation of ET. The estimator of ET should note definitions of both words with the utmost care because the estimation of ET often involves a set of equations that are interrelated. Correct estimates can only be made using proper definitions. The meanings of *potential* or *reference* are frequently discussed at various places in this volume.

In general, the word *potential* was first in time. Potential often meant soil water was not limiting crop growth, and hence the growth rate and ET were not limited by a lack of soil water. *Potential ET* is abbreviated by ET_p. With reference to combination methods, potential meant the evaporating surface was saturated. Crop coefficients were used to convert potential ET to actual ET for a specific plant at the specific time. However, the definition of potential involved a surface that was not well-defined and difficult to visualize.

The use of the word *reference* followed. *Reference crop ET* is indicated by the symbol ET_r. Usually the term reference refers to grass or alfalfa when the height is carefully defined, and the term *basal* is also used and means that the soil surface is dry and that the profile has enough soil water that the rate of ET is not limited. These definitions make it convenient to include cultural corrections, soil water levels, and the frequency of soil surface wetting. These definitions result in an artificial, but useful, calculated ET_r.

5.2.3 *Soil influences*

Soil structurally supports plants, stores almost all of the water used in ET, stores nutrients assimilated by growing plants, and provides an environment conducive to plant growth. Conditions found in soil have a profound impact upon plant growth and, therefore, soil conditions have a very important impact on ET for a specific plant at a specific time.

Satisfactory conditions for one plant may not be those of another plant. For example, the flooded conditions often appropriate for rice are fatal to the well-aerated conditions favorable for alfalfa.

Profile drying

It is axiomatic that dry soil results in reduced plant growth. The whole purpose of irrigation is to make sure that the soil water in the root zone is always within the range leading to the best possible plant growth for the water available.

Plants growing without irrigation must grow with natural rainfall or high water tables as sources and, therefore, may often be growing where soil water is not at optimum levels. The estimation of ET_a then requires a correction for limited soil water.

Water transport from water tables

Movement of water from a water table to the root zone of crops is usually unsaturated. This is in spite of the fact that a zone of near saturation called the capillary fringe always exists above a water table. Upward movement of water is limited by unsaturated hydraulic conductivity which is usually much less than saturated hydraulic conductivity. Unsaturated hydraulic conductivity below the DUL is so small that it is often assumed to be zero. However, movement of soil water will take place inside and to the root zone when portions become dry. Also, water can be supplied to the roots from shallow groundwater. See Chapter 3 for more information.

Fertility

Crop production is obviously increased by fertility. This is particularly true for nitrogen. In general, increased production results in increased ET. See Chapter 6 for relationships between ET and vegetative production.

Salinity

The effects of excessive salinity or alkalinity on crop growth are obvious; crop growth is reduced, and crops exhibit changes in appearance. Chapter 6 which deals with production and ET shows linear relationships between ET and bio-mass production. It is then reasonable that excess salts would reduce ET. Hoffman (1985) shows that ET is linearly related to the leaching fraction when saline water is applied to a number of crops.

5.3 Soil water and ET estimates

The methods presented for estimating reference and crop ET assume soil water is in ample supply. In order to determine actual ET, the level of soil water must be considered. Obviously, there must be water in the soil to supply water for transpiration. As the soil dries out, the rate of ET and the rate of growth are slowed. There are several hypotheses on the effect of limited soil water on plant activity and ET rate.

The depth over which plants extract most of their water can show great differences. As defined in Chapter 3, PESW (available soil water) is simply the amount of water, in mm water per meter soil depth, between the soil water content at the DUL (field capacity) and the soil water content at the LL (wilting point). In general, PESW (available soil water) is about 200 mm/m for heavy textured soils, 140 mm/m for medium textured soils, and 60 mm/m for course textured soils.

Denmead and Shaw (1962) presented some of the first information on the effects of soil water deficits on the water use of plants. They considered mainly reductions in transpiration rates with decreasing soil water levels, without considering the plant and stomatal resistances which affect the transpiration process as shown later by Ritchie (1974).

Available soil water index ASI

Stress indices of various types have been used. Simple indices incorporating supply and demand terms have been developed. One such index is the *available soil water index*, ASI, which indicates the part of a month when available soil water is adequate for meeting full plant water requirements.

$$ASI = \frac{(Water\,supplied) - (Water\,remaining\,when\,\mathrm{ET}_a < \mathrm{ET}_m)}{Monthly\,\mathrm{ET}_m} \tag{5.1}$$

Methods for using ASI are given by Doorenbos and Kassam (1979) as adapted from Rijtema and Aboukhaled (1975).

Classic models of limited soil water

Hillel (1980) shows a sketch of idealized models relating plant activity, and hence ET, as a result of limited soil water within the root zone.

The classical hypothesis illustrated by Hillel (1980) about soil water availability refers to plant activity. Certainly plant activity can include transpiration, and hence ET. The following is intended to show different algorithms used to reduce ET because of limited soil water. There is no reason to expect that relationships between $\mathrm{ET_a}$ and $\mathrm{ET_r}$ or $\mathrm{ET_p}$ as related to soil water be straight lines. Logarithmic, exponential, and other curvilinear relationships have been used.

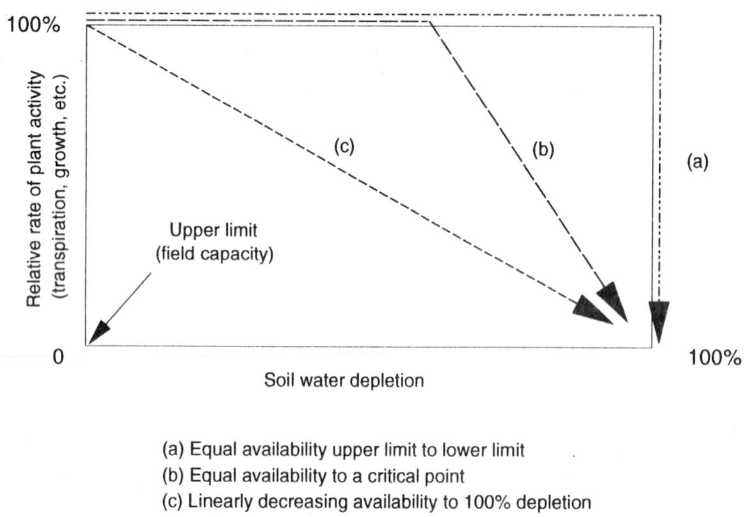

Figure 5.1
Classical hypotheses, soil water availability
from Hillel (1980)

References indicating that the rate of ET for a specific crop at a specific time depends upon soil water are so numerous that the need for specific citations is minimized. There are numerous procedures for correcting the rate of ET for limited soil water. All methods of estimating ET involve regression equations or other empirically determined constants. Therefore, changes in any portion of calculations may influence the final results.

The procedures for making limited soil water corrections are so diverse and often interrelated that it is difficult to make consistent classifications. The following is an attempt to provide an organized presentation of methods of making limited soil water corrections on ET determinations.

5.3.1 *Water depletion & field conditions*

Many practical issues dealing with soil water are not always obvious. *Soil water depletion*, D, is the amount or ratio of *plant extractable soil water*, PESW, or *available soil water*, AW, depleted from soil. Soil water depletion can be negative and over 100% under field conditions. Soil water depletion may refer to the amount of soil water removed from a location within the soil profile, a specific layer within the soil profile, or within the entire root zone. The starting point is the *drained upper limit* (or field capacity) reference point.

Soil water between the DUL and saturation results in a negative depletion, and the soil water in this range will drain by gravity to a lower region. Soil water below the DUL will not drain because the hydraulic conductivity is very small, and the definition of DUL indicates this fact. Soil water may be below the LL especially near the surface of the ground surface. This results in a depletion of over 100%. Depletions over 100% seem illogical but are very possible.

Depletion

The depletion D is defined below. The soil water reference points are discussed in Chapter 3.

$$D = DUL - SW \tag{5.2}$$

where D is depletion, DUL is the drained upper limit (field capacity), and SW is the soil water at a particular time. All are in mm; however, points like the DUL may be shown as percentages of soil water. The equation applies to any location in the soil profile or an entire root zone.

Depletion ratio

$$D_r = \frac{DUL - SW}{DUL - LL} \tag{5.3}$$

where the denominator DUL-LL is called the PESW (available soil water), LL is the lower limit in mm, and now D_r is a decimal. Sometimes D_r is multiplied by 100 and used as a percentage. Both D and D_r are used. An illustration of depletion related to the entire range of soil water follows.

PESWact (existing available soil water)

At times it is convenient to characterize soil water level by the component of depletion.

$$PESW_{act} = \left[\frac{SW - LL}{DUL - LL} \right] \tag{5.4}$$

where SW is the actual soil moisture, DUL is the drained upper limit of soil moisture, and LL is the lower limit of soil moisture all in mm. $PESW_{act}$ is used as a decimal and converted to a percent.

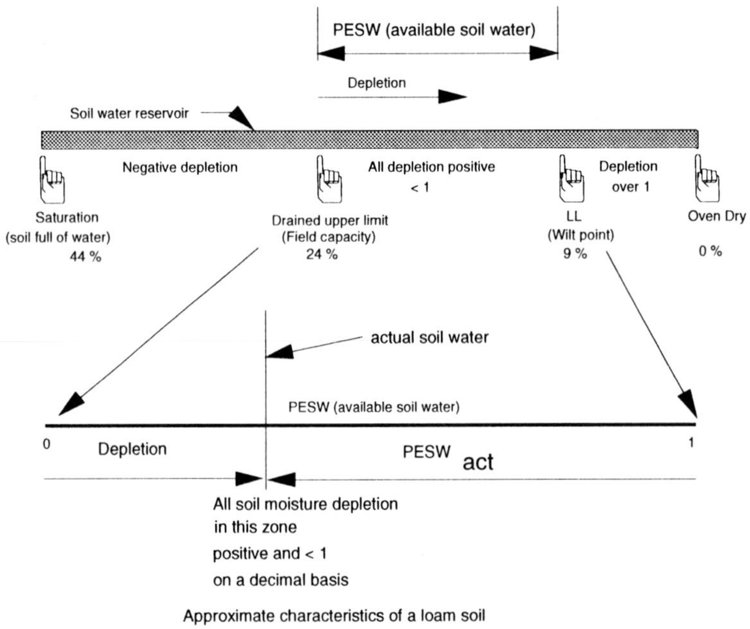

Figure 5.2
Sketch showing soil water depletion
soil water is in volume units

5.3.2 *Soil profile subdivisions*

It is obvious that soil water varies throughout a root zone or other soil profile. From a theoretical standpoint the profile may be divided into zones for accounting purposes. Nieber and Patwardhan (1989) divide the root zone into seven layers. Williams et al. (1985) use a maximum of ten intervals. European soil water accounting models commonly vary from a low of one with two, three, or four level models being used (Andersson and Harding, 1991). The European methods often use the layers to represent amounts of PESW (available soil water), not specific zones in the soil profile. This permits simple convenient corrections for limited soil water to the ET_a rate.

5.3.3 *Zone of ET removal*

Any irrigation scheduling and hydrologic models require the estimation of water removal by depth within the root zone. de Jong and Cameron (1979) make the statement that water removal by ET is proportional to the density of roots in any particular portion of root zone under consideration.

Simple percentage rule of thumb

For many years a very simple rule of thumb regarding the average removal of soil water by portions of the root zone has been used for practical irrigation scheduling. The root zone is divided into quarters with the assumption being made that the soil water removal by quarters from the surface downward is as follows:

1st quarter	40%
2nd quarter	30%
3rd quarter	20%
4th quarter	10%

The origin of this rule of thumb is not known precisely but it has been used on a wide spread basis. This simple rule of thumb was apparently developed by Shockley in the mid 1950's and is discussed by Hansen et al. (1980) and the Irrigation Association (Pair et al., 1983). Hansen et al. (1980) add the qualification that the rule is for average conditions, under irrigation, and in arid regions. Sometimes the qualification is added that there should be *salt free soil water in ample supply*. These percentages are very general and may vary considerably but are, nevertheless, very useful because of their simplicity and ease of use. Numbers which are in even multiples of 10 are easy to remember and, therefore, quite practical in application.

Baier and Robertson model

Baier and Robertson (1966) presented a method of allocating ET_a from different layers in the root zone. The method is known as the *versatile soil water budget*. The equation follows.

$$AE_i = \sum_{j=1}^{n} \{k_j \frac{S'_j(i-1)}{S_j} Z_j\, PE_i \exp[-\omega(PE_i - \overline{PE})]\} \quad (5.5)$$

where AE_i is ET_a on the day which ends on the morning of day i+1, the summation is from zone j = 1 to j = n, $S'_{j(i-1)}$ is soil water in the j th zone on day i - 1, S_j is the available soil water in zone j, Z_j represents the type of soil dryness curve, PE_i is ET_p on day i, ω is an adjustment factor for the rate of ET, and $\overline{PE}$ is the average PE for the month or season

Williams et al. model

Williams et al. (1985) describe the following algorithm to be used to assign ET to different portions of the root zone. The method has been used in a study of effective rainfall by Patwardhan et al. (1990). The method follows:

$$u_{\pi} = \frac{ET_a}{1-\exp(-\Lambda)}\left\{1-\exp\left[-\Lambda\left(\frac{RD_i}{RZ}\right)\right]\right\} - \sum_{i=1}^{i-1} u_i \tag{5.6}$$

where Λ is called a *use rate depth parameter*, u_{pi} is the potential water use from layer *i*, RD_i is the root depth in layer ***i***, and RZ is the total root depth. Both RD_i and RZ are in mm, but fractions may also be used. The use of the word *potential* is unclear, and it appears to be ignored in practice. The recommended value of Λ is 3.065 which is said to represent a root zone where 30% of the total water use comes from the top 10% of the root zone. Other values of Λ can be selected for other distributions of water consumption by depth. The summation term represents the sum of ET above the layer under consideration.

5.3.4 *Measured root density*

Root density is known to change with crop development, soil characteristics, soil water amounts, and other crop environmental conditions. The use of the preceding rule of thumb involving the percentages 40, 30, 20, 10 and other algorithms is so convenient that it is of interest to see how measurements of root density and ET removal zone compare with the rule of thumb.

Table 5.1 shows that the simple 40, 30, 20 and 10 percent rule of thumb is reasonable with expected variations. Apparently root density and zone of ET removal are similar.

5.3.5 *Linear relationships*

Linear relationships may be a straight line between $ET_a=ET_m$ at a depletion ratio of zero to $ET_a=0$ at a depletion ratio of one, or a straight line going to zero after an interval where $ET_a=ET_m$. Figure 5.1 is credited to Hillel (1980). The straight line, option (b) with offset, from Hillel, may be the most logical and simple relationship used. Neibling et al. (1977) in summarizing other references indicate that the offset is a depletion ratio of 0.3. See Section 4.3 for a discussion of ET_a and ET_m.

5.3.6 *Step soil water corrections based on depletion*

Most relationships used to relate ET to soil water are continuous, at least in part. Step corrections obviously are not continuous mathematically. The following step corrections are described by Andersson and Harding (1991). Calder et al. (1983) explain the same procedure and describe the method as the *layer soil water abstraction model.*

Table 5.1
Selected crop root density and zone of ET removal values
(seasonal data only)

Crop Description	quarter Top	2nd	3rd	4th	units	reference
Rule of Thumb	40	30	20	10	% ET removal	Hansen et al 1980
rape, India	40	27	21	12	% ET removal	Raja & Bishnoi 1990
dry conditions	30	28	24	18		
more moist conditions	45	27	19	9		
cotton, Australia, drip & furrow	38	30	19	13	% root density	Constable et al 1990
wheat after rice, India					% root density	Chaudhury et al 1991
0.6 irr to pan E ratio	49	32	12	6		
0.9 irr to pan E ratio	46	32	15	7		
1.2 irr to pan E ratio	48	32	14	6		

Table 5.2
Step correction, ET_a vs depletion

	Depletion interval
ETa = ETm	1
ETa = ETm/2	2
ETa = ETm/4	3
ETa = 0.0	4

Calder et al. (1983)
Andersson and Harding (1991)

ET_a is actual ET and ET_m is maximum ET for a specific plant without a correction for depleted soil water.

The procedure for the step method is first to divide the total soil water into four intervals. These intervals do not have any relationship to specific geometric portions of the root zone. Rather they are critical points in soil water depletion in the total root zone that indicate significant drouth effects on plants. Some authorities recommend making the top two intervals be equal. Andersson and Harding (1991) then recommend selecting the four values by optimization such that the standard deviation between modeled and measured soil water depletion is minimized. They indicate that a simple one level model with linear reduction in ET_a related to depletion gave better results than did the model described above.

5.3.7 *Natural logarithmic relationships*

Jensen et al. (1971) first used the following algorithm to determine a factor for use with reference or potential ET. The model cannot be calibrated but has been used with alfalfa-based ET estimates and is a recommended part of that procedure. A common approach for considering the effects of limited soil water on crop ET is that of adjusting the crop coefficient, as suggested by Jensen et al. (1971). The form of the adjustment is the same as Wright's (1982) method for adjusting the basal crop coefficients.

K_{cc} is the adjusted crop coefficient when soil water is limiting, K_{cb} is the basal crop coefficient for non-limiting soil water conditions, K_a is a soil water stress coefficient which accounts for the effects of limited soil water, and K_s is a coefficient used to adjust for increased evaporation from wet soil immediately after rain or irrigation, and %PESW is the percentage of available soil water. See Section 5.6.2 for more discussion of K and its subscripts. The soil water correction follows:

$$K_a = \frac{\ln(\%PESW + 1)}{\ln(101)} \tag{5.7}$$

where K_a is a coefficient intended to be used as a multiplier with reference or potential ET. The %PESW has been defined previously and is the percent of *plant extractable* (available) *soil water*. K_a is used in conjunction with a correction for a wet soil surface. K_a goes to zero when %PESW = 0 and is 1 when the %PESW equals 100. See Section 5.6.2 for more information.

The model is very convenient and easy to use but suffers because it cannot be changed to accommodate different crops and situations. The model appears to be suited for irrigated conditions.

5.3.8 *Other curvilinear relationships*

Boonyatharokol and Walker (1979) propose the following two parameter general models with the capability of responding to various crops and situations.

$$SW = [1 - (D_r)^m]^n \tag{5.8}$$

where SW is the decimal correction to ET because of limited soil water, and D_r is defined above, and D_r is now a ratio. The powers *m* and *n* are empirical constants related to climate, soil, and plant factors. They present regression equations that can be used to estimate the values of *m* and *n* from saturated hydraulic conductivity and ET_{max}.

Burman (personal communication, 1990) indicates that *m* and *n* vary from 1.6 to 2.0 for high altitude conditions in Colorado with non-irrigated vegetation growing after normal irrigation has stopped.

An example of a correction involving an exponential function is credited to Woodhead by Calder et al. (1983). The authors of this volume do not have access to the publication of Woodhead and the units involved are unknown. In some respects the algorithm may be considered as a composite function which is described below. This is because the relationship is a combination of a zone of constant ET_a being equal to ET_{max} which is followed by the exponential function.

$$ET_a = ET_{max} 1.9 \exp(-0.6523\ D/offset)\ for\ D > offset \tag{5.9}$$

$$ET_a = ET_p for\ D < offset$$

where offset is to be determined by optimization.

5.3.9 *Composite and other relationships*

Calder et al. (1983) show a drawing representing a zone of ET_a being equal to ET_p plus an exponential or decay relationship as a function of mm depletion. The reduction is indirect because it is used to change the value of r_c in the Penman-Monteith equation.

The following illustration shows selected empirical relationships used to reduce ET because of soil water depletion.

The previous relationships all assume that the ratio of ET_a to ET_{max} as related to soil moisture is the same for all crops and at all times in the growing season.

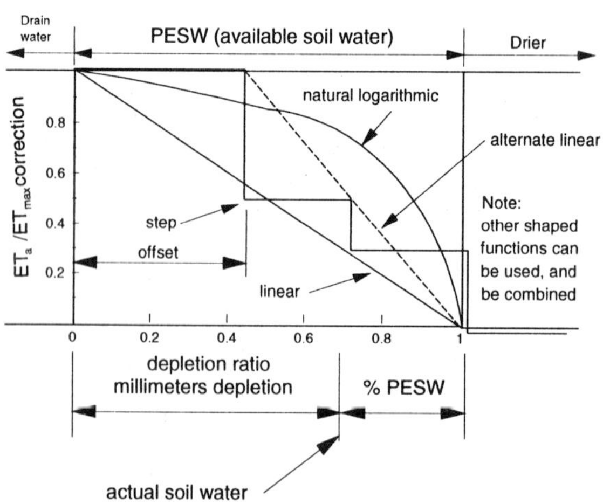

Figure 5.3
Algorithms for ET adjustment for dry soil

The relationship between ET_a and ET_{max} cannot be expected to be the same for all crops at all times. The previous discussion certainly may give the reader that impression. The following material taken from Doorenbos and Kassam (1979) is an example of the effects of different crops and the rate of ET_{max} as well.

Table 5.3
Crop groups as to sensitivity
for the offset to the relationship between ET_a and ET_{max}
from Doorenbos and Kassam (1979)

group	crops
1	onion, pepper, potato
2	banana, cabbage, grape, pea, tomato
3	alfalfa, bean, citrus, grapefruit, pineapple sunflower, watermelon, wheat
4	cotton, maize, olive, safflower, sorghum soybean, sugarbeet, sugarcane, tobacco

Table 5.4
Table showing the fractional offset to
the relationship between ET_a and ET_{max}
from Doorenbos and Kassam (1979)
the numbers 2 through 10 represent ET_m in mm day^{-1}
the numbers 1 through 4 represent crop sensitivity classes

	2	3	4	5	6	7	8	9	10
1	0.500	0.430	0.350	0.300	0.250	0.225	0.200	0.200	0.175
2	0.675	0.575	0.475	0.400	0.350	0.325	0.280	0.250	0.225
3	0.800	0.700	0.600	0.500	0.450	0.425	0.375	0.350	0.300
4	0.885	0.800	0.700	0.600	0.550	0.500	0.450	0.425	0.400

The first table shows that crops are rated as to sensitivity. The second table shows that the decimal offset varies from a low of 0.175 to a high value of 0.875 depending on the sensitivity of the crop and the ET_m value.

5.4 Crop curves

A crop coefficient is the ratio of crop ET to some reference ET at the same time. At first, potential ET was used in this definition (Jensen, 1974). More recently, reference crop ET has been used. Crop coefficients are derived from experimental data. Reference ET, for an identifiable crop under defined conditions with empirical crop coefficients, provides a means of estimating ET for other crops at progressive stages of growth. A plot of a crop coefficient as a function of time or a heat index is known as a crop curve.

In 1974, ASCE (Jensen) defined a crop coefficient as the ratio of evapotranspiration occurring from a specific crop at a specific stage of growth to potential ET at that time. Thus, the definition of crop coefficient predates the use of the term reference crop ET and implied alfalfa-related coefficients because the ASCE committee defined potential ET in terms of the ET from well-watered alfalfa.

Crop coefficients are usually tabulated for different stages of plant growth and for the season. Crop curves permit determining the magnitude of crop coefficients for a desired time period or interval. A sketch of a general crop curve appears in Figure 5.4.

Polynomial representation

Practitioners need the ability to determine crop coefficient values on any desired date. A convenient method is to represent the curve using a polynomial whose coefficients are determined by multiple correlation.

$$f(P) = intercept + A\,P + B\,P^2 + C\,P^3 \qquad (5.10)$$

where P is the percent of time between planting and full cover, or the number of days after full cover for that part of the curve. The variable P may have other convenient units. Polynomial coefficients are presented at various places in discussions of crop curves.

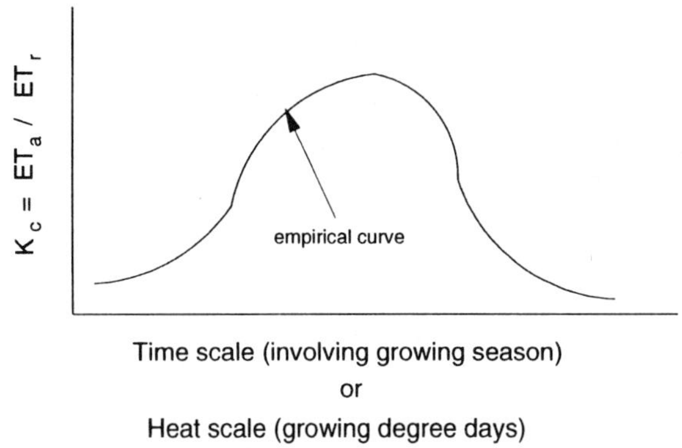

Figure 5.4
Sketch showing a generalized crop curve.

5.4.1 *Basal crop coefficients*

Basal crop coefficients are based upon the definition (Wright 1982; Jensen et al., 1990) of a dry soil surface accompanied with the root zone containing soil water such that transpiration is not reduced. This definition permits the addition of a correction for a wet soil surface which follows a complete irrigation or a significant rain storm. A significant rain is not defined and represents the same issue as is raised in Section 5.8.1. In addition, a correction can also be made for a dry soil profile.

5.4.2 *Average crop coefficients*

Average crop coefficients are higher than basal coefficients and assume that the frequency of rainfall events and irrigations at the location under consideration is typical of the location where data were collected for use in the calibration and the determination of the crop coefficients. Therefore, the use of average coefficients based upon the data of Wright (1982) suggests similar rainfall patterns and numbers of irrigations as were practiced by Wright in the Twin Falls area of southern Idaho, USA.

For comparative purposes both the basal and average definitions of crop curves are illustrated in Figure 5.5.

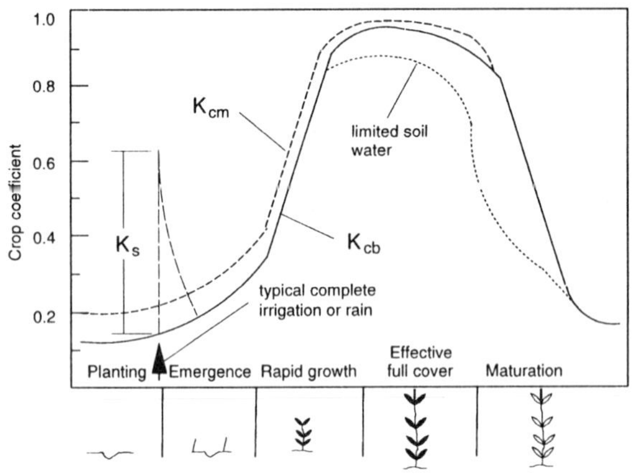

Figure 5.5
Sketch illustrating both average and basal crop curves adapted from Jensen et al. 1990

5.5 Grass-related crop coefficients

Relatively complete listings of grass-related crop coefficients are presented by Doorenbos and Pruitt (1977) and by Doorenbos and Kassam (1979). Doorenbos and Pruitt (1977) presented four methods of calculating reference crop evapotranspiration ET_r, each of which is to be used with the same set of grass-related crop coefficients. The choice of method for calculating ET_r is to be based on the climatic data available and the length of the period for which estimates are being made. Doorenbos and Pruitt (1977) indicate errors can be up to 10 or 25 percent depending upon conditions

Doorenbos and Pruitt (1977) give crop coefficients which vary with crop, stage of crop development, and climate. In addition, the selection of the appropriate crop coefficient depends upon the growing season and time of planting. The climatic conditions considered are combinations of high and low relative humidity, solar radiation, and wind. Climatic conditions are considered since wilting of plants is a possibility when grown under extremely dry windy conditions. Doorenbos and Pruitt state that, in general, as compared to a smooth grass cover, wind will have a more pronounced effect on transpiration of taller plants due to more turbulence above the rougher plant surface. This wind effect is expected to be greatest under dry conditions.

The variety of grass being used as a grass reference by Doorenbos and Pruitt (1977) apparently is not identified. Their definition of reference crop evapotranspiration is *the rate of evapotranspiration from an extensive surface of 8 to 15 cm tall, green grass cover of uniform height, actively growing, completely shading the ground, and not short of water.* There are, of course, differences in ET rates for different grasses, especially when comparing cool season versus warm season varieties (Kneebone, 1979). Alta fescue is a commonly used reference grass and was apparently used by Doorenbos and Pruitt (1977).

calculation procedures

Actual ET is calculated using the simple relationships found in Section 4.3.

$$\mathrm{ET}_m = K_c \mathrm{ET}_r \quad (4.1)$$

$$\mathrm{ET}_a = SW\ \mathrm{ET}_m \quad (4.2)$$

where ET is evapotranspiration in mm day^{-1}, K_c is a dimensionless crop coefficient, and SW is a dimensionless decimal correction for soil water being below the DUL. The subscript *m* means maximum for any crop at a specified time, and '*a*' designates actual.

Growing season dates

Growth stage is obviously important, and Table 5.5 shows growing season criteria recommended by FAO (Doorenbos and Pruitt, 1977). Figure 5.6 shows recommended values of initial stage K_c (Doorenbos and Pruitt, 1977) for different recurrence intervals of irrigations, or significant rainfall, and ET_r magnitudes.

Table 5.6 shows selected lengths of growth stage intervals and total growing season recommended by Doorenbos and Pruitt, 1977. Table 5.7 shows recommended K_c values for forage crops for different humidity and wind conditions (Doorenbos and Pruitt, 1977)

Numerical values of K_c for different crops, climatic conditions, and growth stage conditions are needed to prepare the crop curve. Table 5.8 shows the last two growth stages and has been selected from Doorenbos and Pruitt (1977). The table is similar to one appearing in Burman et al. (1980). The crops considered are nearly the same as those discussed in K_c values for alfalfa-related coefficients.

The initial growth stage crop coefficient is obtained from Figure 5.6. The crop coefficient for the mid-season is merely a straight line drawn between the crop coefficients for the initial and mid-season growth stages. See the rules for constructing a crop curve using the FAO procedures which are found in this section.

Table 5.5
FAO-24 growth stage definitions, Doorenbos and Pruitt (1977)

1 initial stage	germination and early growth when the soil surface is not or hardly covered by the crop (ground-cover < 10%)
2 crop development stage	from the end of initial stage to attainment of effective full ground-cover (ground-cover ≤ 70-80 %)*
3 mid-season stage	from attainment of effective full ground-cover to time of start of maturing as indicated by discoloring of leaves (beans) or leaves falling off (cotton). For some crops this may extend to very near harvest (sugar-beets) unless irrigation is not applied at late season and reduction of ET_a is induced to increase yield and/or quality (sugarcane, cotton, some grains); normally well past the flowering stage of annual crops
4 late season stage	from end of mid-season stage until full maturity or harvest

*start of mid-season stage can be recognized in the field when crop has attained 70 to 80% ground-cover which, however, does not mean that the crop has reached its mature height. Effective full ground-cover refers to cover when K_c is approaching a maximum.

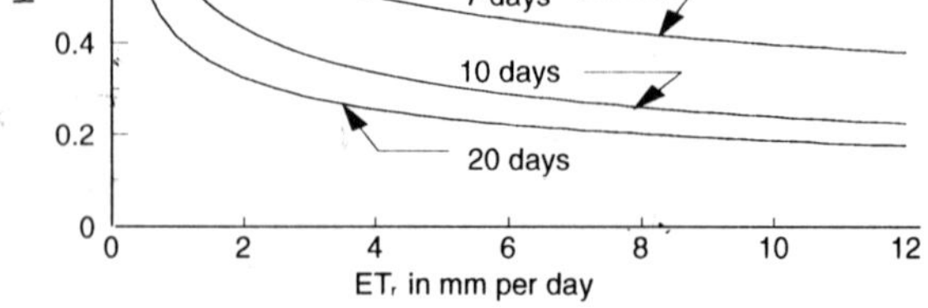

Figure 5.6
K_c for various values of the recurrence interval for irrigation or rainfall for the initial growth stage

Table 5.6
Selected growing season and crop development data
FAO definitions from Doorenbos and Pruitt, 1977

grain, small	spring planting Mediterranean 20/30/60/40 and (150); October-November planting warm winter climates; Pakistan and low deserts 25/35/65/40 and (165).
peas	Cool maritime climates early summer planting /15/25/35/15 and (90); Mediterranean early spring and warm winter desert climates planting 20/25/35/15 and (95); late winter Mediterranean planting 25/30/30/15 and (100).
sugar-beets	Coastal Lebanon, mid November planting 45/75/80/30 and (230); early spring planting Uruguay 30/45/60/45 and (180); late winter planting warm winter desert 35/60/70/40 and (205).
potatoes	Fall planting warm winter desert climates 25/30/30/20 and (105); late winter planting arid and semi-arid climates and late spring-early summer planting continental climate 25/30/45/30 and (130); early-mid spring planting central Europe 30/35/50/30 and (145); slow emergence may increase length of initial period by 15 days during cold spring.
field corn (maize)	Spring planting east African highlands 30/50/60/40 and (180); late cool season planting, warm desert climates 25/40/45/30 and (140); June planting sub-humid Nigeria, early October India 20/35/40/30 and (125); early April planting southern Spain 30/40/50/30 and (150).
sweet corn (maize)	Philippines, early March planting (late dry season) 20/20/30/10 and (80); late cool season planting desert climates 20/30/30/10 and (90); early cool season planting 20/30/50/10 and (110).
beans (dry)	Continental climates late spring planting 20/30/40/20 and (110); June planting Central California and West Pakistan 15/25/35/20 and (95); longer season varieties 15/25/50/20 and (110).
wheat	FAO-24 says to see barley, therefore, barley follows: Also wheat and oats; varies widely with variety; wheat central India November planting 15/25/50/30 and (120); early spring sowing, semi-arid, 35°-45° latitudes and November planting Rep. of Korea 20/25/60/30 and (135); wheat sown in July in East African highlands at 2500m altitude and Rep. of Korea 15/30/65/40 and (150).

15/25/50/20 and (110) stand respectively for initial, crop development, mid-season and late season crop development stages in days and (110) for the total growing season from planting to harvest in days.

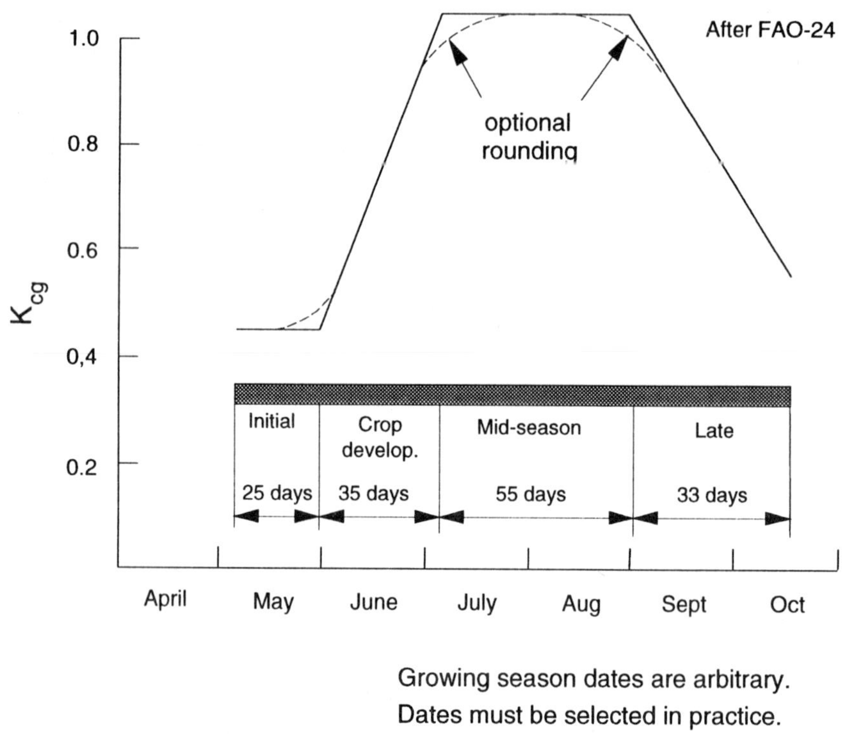

Figure 5.7
Example of an FAO-24 crop curve
days shown are assumed
(days need to be determined for desired crop)

A seven-step procedure for development of crop curves has been recommended by Doorenbos and Pruitt (1977):

1. The planting date is first defined. This requires information on local practices.

2. The lengths of the total growing season and of various developmental stages need to be defined. These are the stages for which crop coefficients are tabulated. Doorenbos and Pruitt (1977) divide the season into four developmental stages. The initial stage extends from germination through the early growth period until the ground cover is no more than 10 percent. The development stage is from the end of the initial stage through the attainment of effective full cover of about 70 to 80 percent. The mid-

season stage extends from the development stage through the time of the start of maturity, as indicated by discoloring or dropping of leaves. The late season stage is the final stage and includes the period through which full maturity or harvest occurs.

3. The crop coefficient for the initial stage is plotted, and a horizontal line, extending from the planting date to the end of the initial stage, is drawn through this value.

4. The crop coefficient for the mid-season stage is plotted, and a horizontal line, extending from the beginning to the end of the mid-season stage, is drawn through this value.

5. The crop coefficient for the time of full maturity is plotted at the end of the growing season, that is, at the time of full maturity.

6. A straight line is drawn from the end of the line passing through the initial stage crop coefficient and the beginning of the line passing through the mid-season crop coefficient. Another straight line is drawn from the end of the line passing through the mid-season crop coefficient to the crop coefficient plotted at the end of the growing season.

7. Although it looks better but generally doesn't affect the accuracy very much, the straight lines can now be replaced by a smooth curve if desired, which is defined as the crop curve.

Figure 5.7 shows a crop curve constructed by the preceding methods using assumed values. See Section 8.5.1 for a numerical example.

Table 5.7
Crop coefficient values K_c
for alfalfa, clover, grass legumes and pasture
FAO-24 grass definition
from Doorenbos and Pruitt (1977)

	K_c	alfalfa	grass hay	clover grass legumes	pasture
humid, with (light to moderate wind)	mean	0.85	0.80	1.00	0.95
	peak	1.05	1.05	1.05	1.05
	low	0.50	0.60	0.55	0.55
dry, with (light to moderate wind)	mean	0.95	0.90	1.05	1.00
	peak	1.15	1.10	1.15	1.10
	low	0.40	0.55	0.55	0.50
no humidity class (strong wind)	mean	1.05	1.00	1.10	1.05
	peak	1.25	1.15	1.20	1.15
	low	0.30	0.50	0.55	0.50

Table 5.8
Crop coefficient values for different crops for different climatic and growth stages selected from Doorenbos and Pruitt (1977)

			humidity			
	crop		RH_{min} > 70%		RH_{min} < 20	
crop	stage	wind m/s →	0-5	5-8	0-5	5-8
grain	3		1.05	1.10	1.15	1.20
	4		0.30	0.30	0.25	0.25
peas	3		1.05	1.10	1.15	1.20
	4		0.95	1.00	1.05	1.10
sugar beets	3		1.05	1.10	1.15	1.20
	4		0.90	0.95	1.00	1.00
potatoes	3		1.05	1.10	1.15	1.20
	4		0.70	0.70	0.75	0.75
field corn	3		1.05	1.10	1.15	1.20
	4		0.55	0.55	0.60	0.60
sweet corn	3		1.05	1.10	1.15	1.20
	4		0.95	1.00	1.05	1.10
beans (dry)	3		1.05	1.10	1.15	1.20
	4		0.30	0.30	0.25	0.25
wheat	3		1.05	1.10	1.15	1.20
	4		0.25	0.25	0.20	0.20

Table 5.9
Basal crop coefficients for the reference crop grass
FAO-24 definition
from Jensen et al. (1990)
% of days from planting to peak K_c

crop	plant date	days to full cover	10	20	30	40	50	60	70	80	90	100
sorghum	5/17	45	0.12	0.13	0.14	0.16	0.22	0.33	0.50	0.75	1.00	1.07
beans	6/21	43	0.10	0.12	0.16	0.21	0.28	0.39	0.53	0.75	0.98	1.08
tomatoes	4/29	80	0.14	0.15	0.17	0.19	0.22	0.33	0.48	0.71	1.04	1.18
barley	10/31	100	0.18	0.20	0.22	0.24	0.28	0.34	0.47	0.66	0.90	1.07
corn	5/14	52	0.12	0.13	0.15	0.20	0.29	0.45	0.81	0.99	1.08	1.13
sugar beets (late)	6/16	55	0.12	0.13	0.16	0.20	0.29	0.45	0.65	0.87	1.04	1.10
sugar beets (early)	3/25	90	0.14	0.16	0.18	0.22	0.27	0.37	0.53	0.77	1.04	1.10

Table 5.9 contains grass related K_c values for basal conditions. The format presents K_c values for the percent of time from planting to the time of maximum ET and then a separate Table 5.10 shows days from the time of peak ET.

5.6 Alfalfa-related crop coefficients

Alfalfa is preferred over grass as a reference crop by some workers. The main advantage, as discussed by Jensen (1974), is that the aerodynamic characteristics of alfalfa are more like those of most agricultural crops. Also, the greater depth of the plant canopy absorbs more solar radiation than will grass and, finally, the root system is greater in depth than that of grass. However, disadvantages of alfalfa include the fact that grass can be clipped and thus maintained at a near constant height throughout the growing season. Alfalfa, on the other hand, passes through a number of growth and harvest cycles, depending upon the location. Thus, alfalfa does not provide nearly as consistent a reference as does grass during the season. What is being compared is the water use rate from a fully developed alfalfa plant with the water use rates of agricultural crops at various stages of development. The timing of crop maturity is such that fully developed alfalfa does not occur simultaneously with the growth stages of agricultural crops.

Jensen (1974) presents seasonal alfalfa-related crop coefficients for eight agricultural crops. As shown, these include coefficients for alfalfa before full development of the alfalfa canopy (after full canopy development, the coefficients have a magnitude of one). Jensen defines the reference crop as an actively growing well-watered alfalfa at least 30 cm tall. Wright (1982) presents alfalfa-related crop coefficients which he has termed *average* crop coefficients. Average coefficients are suggested for instances when it is not possible to estimate the effects of surface soil wetness, and the timing of irrigations and rainfall is similar to conditions found at Kimberly, Idaho. *Basal* crop coefficients are recommended when it is possible to correct for actual rainfall and irrigation events.

Table 5.10
Basal crop coefficients for the reference crop grass
FAO-24 definition
from Jensen et al. (1990)
days after peak K_c

crop	harv date	days from full cover	10	20	30	40	50	60	70	80	90	100
sorghum	9/13	74	1.08	1.06	1.03	0.99	0.94	0.88	0.79	0.65		
beans	9/18	46	1.12	1.12	1.10	0.71	0.15					
tomatoes	9/24	68	1.24	1.21	1.12	1.03	0.90	0.75	0.58			
barley	5/19	100	1.15	1.17	1.19	1.21	1.19	1.12	0.98	0.75	0.50	0.24
corn	9/20	77	1.17	1.17	1.17	1.14	1.03	0.87	0.67			
sugar beets (late)	11/18	100	1.15	1.16	1.16	1.16	1.15	1.14	1.13	1.12	1.10	1.08
sugar beets (early)	9/20	90	1.13	1.15	1.15	1.14	1.13	1.11	1.08	1.05	1.01	

To use information from Tables 5.9 and 5.10, the estimator must determine the date when the peak K_c occurs. The criteria shown in Table 5.5 may be of assistance. The footnote to the table says *effective full ground-cover refers to cover when K_c is approaching a maximum.* This is a quote from Doorenbos and Pruitt (1977).

Growing season dates

Wright (1982) and Jensen et al. (1990) present detailed dates of various crop growth stages for Kimberly, Idaho but the data is limited to the area cited or similar areas. Governmental agencies often produce management aids which show dates of growth stages for specific areas and adapted crops. Examples are *irrigation guides* published by the US Soil Conservation Service. The use of crop coefficients shown in this section require estimates of growing season dates. While a full definition of growing season dates is desirable, the knowledge of the dates of planting and full cover are essential.

Growing season dates refer to plants and are, therefore, independent of methods and procedures used to estimate ET. However, classifications of divisions of the growing season are arbitrary and will differ with methodology. Recognizing differences in nomenclature, growing season dates are discussed in FAO-24, and the reader is referred to Tables 5.5 and 5.6.

The estimation of growing season dates for ET estimation usually involves the accumulation of dates from research and governmental reports. In addition, much useful knowledge can be obtained from interviews with farmers, agricultural extension personnel, and other knowledgeable people. The final selection of dates, therefore, involves judgment using many sources of information.

5.6.1 *Use of alfalfa-related average crop coefficients*, K_c

Actual ET is calculated using the simple relationships found in Section 4.3.

$$\mathrm{ET}_m = K_c \mathrm{ET}_r \tag{4.1}$$

$$\mathrm{ET}_a = SW\ \mathrm{ET}_m \tag{4.2}$$

where ET is evapotranspiration, and SW is a dimensionless decimal correction to account for reduced soil water content below the DUL in the root zone. The subscripts '*a*' and '*m*' indicate actual and maximum.

Table 5.11
Average crop coefficients, before full cover percent planting to full cover
from Wright (1982) and Jensen et al. (1990)
User needs to identify dates of planting
and full cover for desired crop

	0	10	20	30	40	50	60	70	80	90	100
spring grain	0.20	0.20	0.21	0.26	0.39	0.55	0.66	0.78	0.92	1.00	1.00
peas	0.20	0.20	0.21	0.26	0.36	0.43	0.51	0.62	0.73	0.85	0.93
sugar beets	0.20	0.20	0.21	0.22	0.24	0.27	0.33	0.45	0.60	0.80	1.00
potatoes	0.20	0.20	0.20	0.22	0.31	0.41	0.51	0.62	0.70	0.76	0.78
corn	0.20	0.20	0.20	0.20	0.23	0.32	0.42	0.55	0.70	0.85	0.95
beans	0.20	0.20	0.20	0.26	0.35	0.45	0.55	0.66	0.80	0.90	0.95
winter wheat	0.30	0.30	0.30	0.50	0.75	0.90	0.98	1.00	1.00	1.00	1.00

Table 5.12
Average coefficients, days after full cover
from Wright (1982) and Jensen et al. (1990)
User needs to identify date of full cover for desired crop

	0	10	20	30	40	50	60	70	80	90	100
spring grain	1.00	1.00	1.00	1.00	0.90	0.50	0.30	0.15	0.10		
peas	0.93	0.93	0.70	0.53	0.35	0.20	0.12	0.10			
sugar beets	1.00	1.00	1.00	1.00	0.98	0.94	0.89	0.85	0.80	0.75	0.71
potatoes	0.78	0.78	0.76	0.74	0.71	0.67	0.63	0.59	0.36	0.25	0.20
field corn	0.95	0.96	0.95	0.94	0.90	0.85	0.79	0.74	0.35	0.25	
sweet corn	0.95	0.94	0.93	0.90	0.85	0.75	0.58	0.40	0.20	0.10	
beans	0.95	0.95	0.90	0.67	0.33	0.15	0.10	0.05			
winter wheat	1.00	1.00	1.00	1.00	0.95	0.55	0.25	0.15	0.10		

The SW correction will usually be assumed to be one when average coefficients are used.

K_c values for the forage crops alfalfa and grass follow.

Table 5.13
Average K_c values for forage crops
alfalfa reference crop definition
values by alfalfa cutting
% of time since new growth or harvest
from Wright (1982) and Jensen et al. (1990)

% time	1st	intermediate	last
0	0.55	0.30	0.30
10	0.70	0.40	0.40
20	0.82	0.50	0.50
30	0.91	0.80	0.60
40	0.96	0.96	0.65
50	0.99	0.99	0.63
60	1.00	1.00	0.61
70	1.00	1.00	0.59
80	0.98	0.98	0.57
90	0.96	0.96	0.55
100	0.94	0.94	0.50

Table 5.14
Average K_c values for forage crops
alfalfa reference crop definition
days from planting
from Wright (1982) and Jensen et al. (1990)

crop	0	20	40	60	80	100	120	140	160	180	200
alfalfa	0.45	0.69	0.87	0.88	0.70	0.75	0.88	0.81	0.88	0.71	0.65
seasonal	0.50	0.85	0.85	0.85	0.85	0.85	0.85	0.85	0.85	0.75	0.60
overall average						0.82					
grass	0.60	0.70	0.78	0.78	0.78	0.78	0.78	0.78	0.78	0.76	0.75

Polynomial coefficients for equation 5.10 for average conditions follow.

Table 5.15
Polynomial coefficients for average crop coefficients
alfalfa reference crop definition
Input data from Wright (1982) & Jensen et al. (1990)
% of time planting to full cover

Crop	Interc	A	B	C	r^2
spring grain	0.2127	-0.7195	3.8788	-2.3621	0.9978
peas	0.1997	-0.2065	1.6842	-0.7421	0.9985
sugar beets	0.2032	0.002059	-0.2139	1.0217	0.9982
potatoes	0.2149	-0.6553	2.9650	-1.7405	0.9976
corn	0.2217	-0.6116	1.9668	-0.6041	0.9950
beans	0.2078	-0.4679	2.5618	-1.3403	0.9985
winter wheat	0.2531	0.2777	2.9988	-2.5874	0.9598

Table 5.16
Polynomial coefficients for average crop coefficients
alfalfa reference crop definition
Input data from Wright (1982) & Jensen et al. (1990)
Days after full cover

Crop	Interc	A	B	C	r^2
spring grain	0.9474	0.02079	-0.865E-3	0.585E-5	0.9693
peas	0.9512	-0.00274	-0.000535	5.75E-6	0.9942
sugar beets	0.9927	2.468E-3	-9.04E-5	3.71E-7	0.9968
potatoes	0.7721	1.191E-3	-5.66E-5	-1.65E-7	0.9708
field corn	0.9627	7.99E-4	6.853E-5	-1.69E-6	0.9621
sweet corn	0.9267	6.092E-3	-2.5E-4	8.411E-7	0.9919
beans	0.9439	0.01408	-1.22E-3	1.066E-7	0.9898
winter wheat	0.9389	0.02257	-0.000901	6.019E-6	0.9640

5.6.2 *Use of alfalfa-related basal crop coefficients*, K_{cb}

Wright (1982) presented alfalfa-related *basal* crop coefficients for use when the soil surface is dry, but the root zone water is non-limiting. Although the effort and information required for use of basal coefficients are greater than when using mean coefficients, Wright contends that use of the basal coefficients permits more accurate estimates of daily crop ET. Since they are applied to a dry soil surface, basal coefficients estimate essentially the transpiration component of ET. Crop coefficients for use in the crop ET versus reference ET ratio are calculated by applying two adjustments to the basal coefficients:

$$K_{cc} = K_{cb}K_a + K_s \tag{5.11}$$

$$K_a = \ln(\%PESW + 1)/\ln(101) \tag{5.12}$$

$$\mathrm{ET}_a = K_{cc}\,\mathrm{ET}_{ra} \tag{5.13}$$

where K_{cb} is the *basal crop coefficient*, K_a is a dimensionless coefficient dependent on available soil water, and K_s is a coefficient used to adjust for increased evaporation from wet soil immediately after rain or irrigation. %PESW (available soil water, A_w)is the *percent plant extractable soil water* that occurs in the root zone on the day under consideration (see Section 3.3.1). The value of K_a ranges from one for a well-watered plant to less than one when soil water limits transpiration. The value of K_s ranges from zero for a dry soil surface to greater than zero for a wet soil surface.

Values of K_{cb} for common commercial crops are shown in Tables 5.17 through 5.19.

Table 5.17
Basal values of k_{cb} for various crops
from Wright (1982) and Jensen et al. (1990)
% of time from planting to full cover

crop	0	10	20	30	40	50	60	70	80	90	100
spring grain	0.15	0.15	0.16	0.20	0.25	0.40	0.52	0.65	0.81	0.96	1.00
peas	0.15	0.15	0.16	0.18	0.20	0.29	0.38	0.47	0.65	0.80	0.90
sugar beets	0.15	0.15	0.15	0.15	0.15	0.17	0.20	0.27	0.40	0.70	1.00
potatoes	0.15	0.15	0.15	0.15	0.15	0.20	0.32	0.47	0.62	0.70	0.75
corn	0.15	0.15	0.15	0.16	0.17	0.18	0.25	0.38	0.55	0.74	0.93
beans	0.15	0.15	0.16	0.18	0.22	0.34	0.45	0.60	0.75	0.88	0.92
winter wheat	0.15	0.15	0.15	0.30	0.55	0.80	0.95	1.00	1.00	1.00	1.00

Table 5.18
Basal values of k_{cb} for various crops
from Wright (1982) and Jensen et al. (1990)
days from full cover

crop	0	10	20	30	40	50	60	70	80	90	100
spring grain	1.00	1.00	1.00	1.00	0.90	0.40	0.15	0.07	0.05		
peas	0.90	0.90	0.72	0.50	0.32	0.15	0.07	0.05			
sugar beets	1.00	1.00	1.00	1.00	0.98	0.91	0.85	0.80	0.75	0.70	0.65
potatoes	0.75	0.75	0.73	0.70	0.66	0.63	0.59	0.52	0.20	0.10	0.10
field corn	0.93	0.93	0.93	0.90	0.87	0.83	0.77	0.70	0.30	0.20	0.15
sweet corn	0.93	0.91	0.90	0.88	0.80	0.70	0.50	0.25	0.15		
beans	0.92	0.92	0.86	0.65	0.30	0.10	0.05				
winter wheat	1.00	1.00	1.00	1.00	0.95	0.50	0.20	0.10	0.05		

Table 5.19
Basal values of K_{cb} for alfalfa by cutting
% of time since growth
from Wright (1982) and Jensen et al. (1990)

% time	1st	intermediate	last
0	0.40	0.25	0.25
10	0.50	0.30	0.30
20	0.62	0.40	0.40
30	0.80	0.70	0.50
40	0.90	0.90	0.55
50	0.95	0.95	0.50
60	1.00	1.00	0.40
70	1.00	1.00	0.35
80	0.98	0.98	0.30
90	0.96	0.96	0.27
100	0.94	0.94	0.25

5.6.3 *Wet soil surface*

It is almost always recognized that a wet soil surface may lead to ET near reference or the potential rate (Ritchie, 1972; Wright, 1979). This phenomenon is one of the rules followed in computing ET by the component method. See Section 5.8.1. The following procedure is recommended for calculations involving the above equations which are tailored for alfalfa-related reference crop calculations. The procedure is credited to Wright (1981) and described in detail by ASCE (Jensen et al., 1990) and ASAE (Hatfield and Fuchs, 1990).

$$K_s = (K_1 - K_{cb})\,[1 - \sqrt{t/t_d}\,]\,f_w \tag{5.14}$$

where K_s and K_{cb} have been defined previously. K_1 is the maximum value of K_c that normally occurs after an irrigation or significant rain. K_1 is often assigned the value of 1, (Hill et al., 1983). The letter *t* is time in days, and the subscript *d* means time to a dry soil surface. The time to a dry soil surface primarily depends upon soil characteristics. The symbol f_w is the fraction of soil surface wetted by irrigation. General values of the time needed for soil to dry and portion of wetted soil appear in Table 5.20.

Kincaid and Heermann (1974) show the following simplification for calculating the increased evaporation caused by a wet soil surface for use in western Nebraska:

$$E_{ws} = K_r(0.9 - K_{c'})\mathrm{ET}_r \tag{5.15}$$

where ET_r is their version of reference crop ET, and $K_{c'}$ is their version of a crop coefficient. The time required for the soil to dry was assumed to be three days with K_r having values of 0.8, 0.5, and 0.3. The same system of calculation was described by Burman et al. (1980).

5.7 Estimation of crop ET without formal ET_r

The concept of reference crop ET, or its similar concept of potential ET, may be difficult to apply under some conditions. Reference ET must be carefully defined to be useful. The common reference crops of alfalfa and grass are defined as to kind of crop, surface culture, and crop height. Reference crop ET is carefully defined as used here. However, potential ET is poorly defined and may cause confusion.

Table 5.20
General values of time required for a dry soil surface and amount of ground surface wetted from Hill et al. (1983)

	class	t_d
soil	clay loam	7-11 days
	loam	4-5 days
	sandy	3 days
	type	f_w
irrigation	sprinkler, border every furrow	1.0
	alternate furrow	0.5
	drip	0.25

Reference crop ET also has limitations because the commonly used reference crops, alfalfa and grass, may be quite different botanically as conditions and locations vary (Pochop and Burman, 1987). Another limitation is that alfalfa may not be grown at high altitude locations or under acid soil conditions. For these reasons, estimation of ET_a without the use of a carefully defined reference crop involving ET_r may have advantages in some instances.

5.7.1 *Blaney-Criddle method*

Historically, the most widely applied of all ET equations has probably been the Blaney-Criddle equation. Various versions of the Blaney-Criddle equation have been developed with the original equation appearing in 1950 (Blaney and Criddle, 1950). See Section 4.7.1 for more discussion of the Blaney-Criddle method.

The Blaney-Criddle equation utilizes temperature and day length data. The day length data partially accounts for radiation variations between locations and with time. In addition, coefficients are applied to estimate ET for specific crops.

The Blaney-Criddle method is a simplification of a method proposed by

Blaney and Morin (1942) which related consumptive use to day length, mean air temperature, and humidity. A lack of humidity data led to the Blaney-Criddle method which relates consumptive use to day length and temperature (Blaney and Criddle, 1950 and 1962). The Blaney-Criddle method has seen a considerable amount of modification during its 45-year existence. The following discussion is intended to emphasize the point that the exact version used must be carefully identified to avoid misinterpretation of results. Only the FAO-24 version of the Blaney-Criddle method involves the use of a well defined grass-based reference crop.

The first version of the Blaney-Criddle method utilized a seasonal coefficient, K_c, known as an *empirical seasonal coefficient* which varied with the crop but was constant for an entire growing season (Blaney and Criddle, 1950). It soon became evident that the coefficient should not be constant for an entire season but should vary from month to month. Blaney and Criddle (1962) present the modification of using monthly coefficients. They show a compilation of monthly coefficients called *monthly consumptive use crop coefficients.*

The simple Blaney-Criddle method, as used historically in many areas, is shown below.

$$ET_a = 25.4 K_c' F \tag{5.16}$$

where ET is *seasonal evapotranspiration* in mm, K_c' is an *empirical seasonal consumptive use coefficient*, and F is the sum of monthly f values. In at least one place, Blaney and Criddle (1962) refer to K_c as a *crop coefficient.*

$$f = \frac{(1.8T + 32)p}{100} \tag{5.17}$$

where f is a *monthly consumptive use factor*, T is average *air temperature* in °C, and p is the *percent daylight hours* of a month with respect to the entire year.

The use of the seasonal K_c' did not permit monthly estimates of ET_a so provision was made for monthly calculations.

$$u = 25.4\, k_c' f \tag{5.18}$$

where u is monthly consumptive use in mm, and k_c' is called the *monthly consumptive use coefficient*. This coefficient should not be confused with the crop coefficient used with the usual definition potential or reference crop ET. Jensen (1974) and Jensen et al. (1990) show that the Blaney-Criddle consumptive use coefficient has meteorological or other influences present.

The argument can be made that *tp*/100 represents an approximation of ET_p or ET_r and, therefore, the Blaney-Criddle method is a form of procedures where ET_a estimates are made using a reference crop. However, the argument is very controversial but, nevertheless, the Blaney-Criddle method remains a very widely used method of estimating ET, and many practitioners refer to the *empirical monthly consumptive coefficients* as *crop coefficients.*

5.7.2 *SCS Blaney-Criddle method*

In 1967 the US Soil Conservation Service published TR-21 *Irrigation Water Requirements* (USDA, 1970). This publication is based on a study of ET from many locations. The locations used in calibrating the SCS TR-21 version of the Blaney-Criddle method appear in a map published by Nieber and Patwardhan (1989). These locations should be helpful in assessing the suitability of the method for making ET estimates in a specific location. In the past, coefficients were often greater than one in the spring and fall, and less than one in mid-summer. In an attempt to eliminate this loop effect, a compound *k* factor was introduced. The final *k* used in calculations of ET is a product of a k_c *reflecting the growth stage of the crop*, and k_t *a climatic coefficient which is related to the mean air temperature.* The coefficient k_t was apparently determined using a regression equation developed from ET and air temperature data, assuming that k_c is equal to one. When using the TR-21 version of the Blaney-Criddle method, it is essential that consistent k_t values be used. The SCS TR-21 Blaney-Criddle method follows:

$$ET_a = \hat{k}_c k_t f \qquad \textit{monthly} \tag{5.19}$$

Now the compound k_c' is defined as follows:

$$k_c' = \hat{k}_c\, k_t \tag{5.20}$$

The $\hat{k}_c$ is defined as the Blaney-Criddle *monthly consumptive use coefficient* for the SCS TR-21 version, and the k_t coefficient was defined as a *climatic coefficient* which was related to the mean air temperature as:

$$k_t = 0.0311T + 0.24 \tag{5.21}$$

Pochop et al. (1984) present an alternative function for k_t for alfalfa grown under a variety of altitudes. The expression for k_t follows:

$$k_t = 0.0254\,T + 0.376 \tag{5.22}$$

where variables in the above equations have previously been defined.

As with most ET formulas, better estimates can be achieved with the SCS Blaney-Criddle version if some local calibration is performed. A major problem with the SCS Blaney-Criddle is that a standard method for calibration has not been adopted. Most calibrations have involved calculating new crop growth stage coefficients for the desired crop and local area as done, for example, by Borrelli et al. (1981) for western turf grass. The lack of appropriate calibrations may be one reason that many users have found ET estimates from the SCS Blaney-Criddle to be less than satisfactory. This may also be one of the contributing factors to many efforts to develop additional equations rather than accepting the Blaney-Criddle method.

The k'_c values for earlier versions of the Blaney-Criddle method should not be used with the SCS TR-21 version.

The $\hat{k}_c$ coefficient was defined as a coefficient reflecting the growth stage of the plant and was obtained from growth stage curves. Although k_t and $\hat{k}_c$ are defined as coefficients representing climatic and crop characteristics, respectively, both have been recognized as including other factors. For example, Jensen (1974) states that the crop growth stage coefficient also contains meteorological components. On the other hand, the climatic factor may also contain a plant component

The approach recommended herein for calibrating the SCS Blaney-Criddle formula is first to calibrate the climatic factor by regressing measured ET against temperature. If this step is omitted and only the crop growth stage coefficients are calibrated, the resulting calibration may produce an inverted seasonal curve for the crop growth stage coefficients (Borrelli et al., 1981). If the climatic coefficient is re-calibrated prior to calibration of the crop growth stage coefficients, then the seasonal curve of the crop growth stage coefficients will be the expected bell-shaped curve (Hill et al., 1989).

5.7.3 *Penman-Monteith method*

Thompson et al. (1981) describe the use of direct calculations of plant ET for a network of stations in the United Kingdom using the Monteith version of the Penman equation. The program is known as MORECS and is sponsored by the British Meteorological Office. The performance of the MORECS model is discussed by Andersson and Harding (1991). The implementation of Thompson et al. (1981) makes weekly and monthly estimates of ET and soil water deficits for Great Britain using daily synoptic weather data. Traditional commercial crops such as potatoes and barley are included. In addition, the non-commercial plant surfaces of both coniferous and deciduous trees and lawns are discussed. The following four equations describing the Penman-Monteith method are similar to those found in Chapter 4 but have been re-stated in the same form as those of Thompson et al. (1981). The units used here differ from those of Thompson et al. (1981).

The plant resistance term r_c is defined as representing the entire plant surface at any given time, but with the additional stipulation that soil water does not limit plant growth or ET. Two procedures for adjusting r_c for dry soil water are presented at the end of this section.

$$\lambda ET = \frac{\Delta H + \rho C_p Q}{\Delta + \gamma^*} \tag{5.23}$$

$$H = R_n + G \tag{5.24}$$

$$Q = \frac{e^o[T_z] - e_z}{r_a} \tag{5.25}$$

$$\gamma^* = \gamma\left[1 + \frac{r_c}{r_a}\right] \tag{5.26}$$

where Δ is the *slope of the saturation vapor pressure-temperature curve*, kPa K^{-1} and γ is the *psychrometer coefficient* kPa K^{-1}. See Chapter 8 for a detailed unit analysis.

Table 5.21
Recommended values of $r_{c(min)}$ for various plants, soil water not limiting
r_c in sec m^{-1}

community	Jan	Feb	Mar	Apr	May	June	July	Aug	Sept	Oct	Nov	Dec	r_a	reference
grass & riparian	80	80	60	50	40	60	60	70	70	70	80	80	2	Thom. Et al 81
conifers	70	70	70	70	70	70	70	70	70	70	70	70	2	Thom. Et al 81
upland	120	120	120	100	100	100	100	100	100	120	120	120	2	Thom. Et al 81
pine forest, Engla				114	106	98	125	150	120				1	Szeicz et al 69
pine forest, Calif				114	106	98	125	150	120				1	Szeicz et al 69
potatoes, Englan				130	126	113	45	60	110				1	Szeicz et al 69
potatoes, Calif				130	125	113	67	90	110				1	Szeicz et al 69
lucerne, England				70	56	64	35	30	60				1	Szeicz et al 69
lucerne, Calif				70	56	64	70	60	60				1	Szeicz et al 69

Methods of adjusting r_s for limited soil water follow. The first is an equation suggested by Monteith (1965).

$$r_s = \left[\frac{PE}{AE} - 1\right]\left[\frac{\Delta}{\gamma} + 1\right] r_a + \frac{PE}{AE} r_{c(\min)} \tag{5.27}$$

where Δ, γ, and r_a have been previously defined. PE and AE represent potential and actual evapotranspiration.

The following expression was suggested by Russell (1980) and applies to a two level soil water model.

$$r_s = r_{c(\min)}\left[3.5\left(1 - \frac{BL}{BMAX}\right) + \exp\left(\frac{BMAX}{BL} - 1\right)\right] \tag{5.28}$$

where r_a has been previously defined. BMAX is the maximum soil water capacity of the lower of two levels, and BL is the observed soil water magnitude of the lower level. Both are usually expressed in mm. The following table shows selected resistances observed by various researchers.

Table 5.22
Selected values of r_c observed by various workers
r_c in sec m^{-1}

plant community	meaning	high r_c	low r_c	r_a	water conditions	reference
cereals	surface		40	2	not limiting	Thompson et al 81, England
potatoes & beets	surface		40	2	not limiting	Thompson et al 81, England
deciduous trees	surface		80	2	not limiting	Thompson et al 81, England
bare soil	surface		100	2	not limiting	Thompson et al 81, England
water	surface		0	2	not limiting	Thompson et al 81, England
barley	crop	200	15	1	incl dry soil	Russell, 1980, England
pasture	crop	175	0	1	incl dry soil	Russell, 1980, England
sorghum	leaf	1127	381	2	incl dry soil	Szeicz et al 1973, Texas
hardwood forest	canopy	7000	300	?	incl dry soil	Singh & Szeicz 1980, Quebec
salt cedar	stomatal	740	30	3	incl dry soil	Hughes 1972, Mid Rio Grande
barley	crop	230	30	2	incl dry soil	Monteith et al 1965, England
grassland	surface	280	150	2	inc dry soil	Dunin et al 1978, Australia
grassland	surface	93	35	2	inc dry soil	Dunin et al 1978, Australia
potatoes	surface	22	7	4	soil drying	Nkemdirm 1976, Canada
soybean	stomatal	550	100	P	40% depl	Brady et al 1975, Kansas USA
wheat	surface	1000	100	2	incl dry soil	Wallace et al 1981 India
dhal	surface	1000	78	2	incl dry soil	Wallace et al 1981, India
lentil	surface	1000	59	2	incl dry soil	Wallace et al 1981, India
linseed	surface	1000	250	2	includes dry soil	Wallace et al 1981, India
mixed crop	surface	1000	91	2	incl dry soil	Wallace et al 1981, India
bare soil	surface	10000	1666	2	incl dry soil	Wallace et al 1981, India

5.7.4 *Flow chart for direct estimation of ET*

The following flow chart is a visual attempt to summarize the three methods of making direct estimates of ET. The flow chart is also an attempt to summarize what can be a very complex computational process.

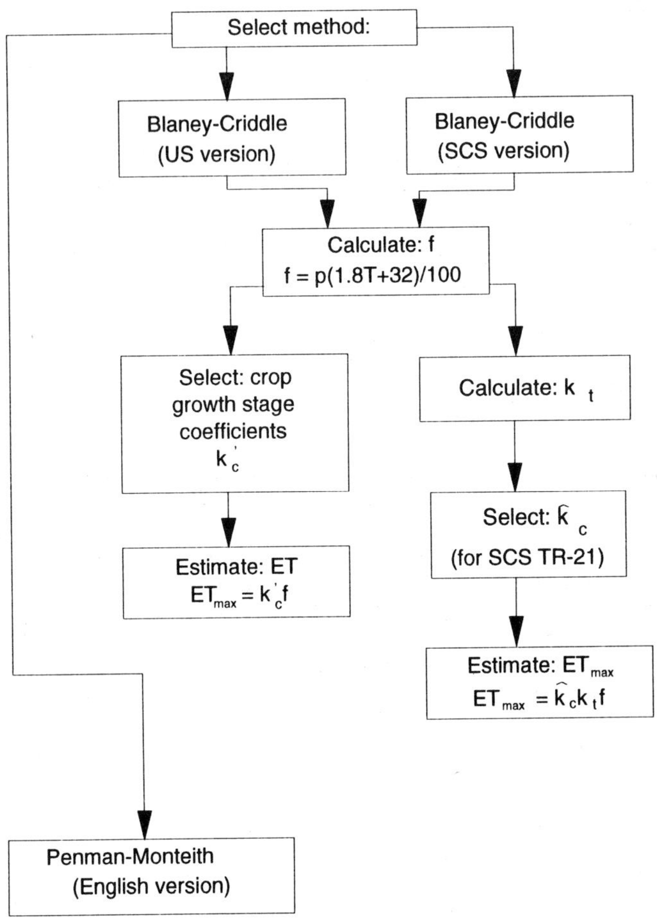

Figure 5.8
Flow chart for direct ET estimates.
for the English version of the Penman-Monteith method
see Section 5.7.3

5.8 Separate estimation of E_s and T

The separate estimation of both *soil evaporation*, E_s, and *transpiration*, T, is a significantly different approach to the estimation of ET. Of course, ET is a simple sum of the two. This approach has the advantage of automatically incorporating the dramatic correction to soil evaporation caused by a wet soil surface. Using this approach, it seems to be more difficult to introduce cultural factors than when making direct estimates of ET. The cultural factors include the type of evaporating surface and aerodynamically characteristics, such as a rough crop like alfalfa. Wet soil evaporation is insignificant when the plants or crop shade the soil surface.

Often hydrologists need to predict ET from fields both before and after a commercial crop is produced. During these times the surface of the field may be bare soil, stubble, or crop residue from the previous crop, or mixtures of all three. The prediction of ET, or E, under these circumstances may involve different procedures than those used to predict ET for commercial crops. The incorporation of these procedures will involve adaptation and somewhat arbitrary decisions. Figure 5.9 is a flow diagram showing typical calculations.

5.8.1 *Soil evaporation*

The fact that soil evaporation occurs under at least two different regimes is universally accepted. This is true even when vegetation or crop residue is present. References by Ritchie (1972) and Philip (1957) are classics.

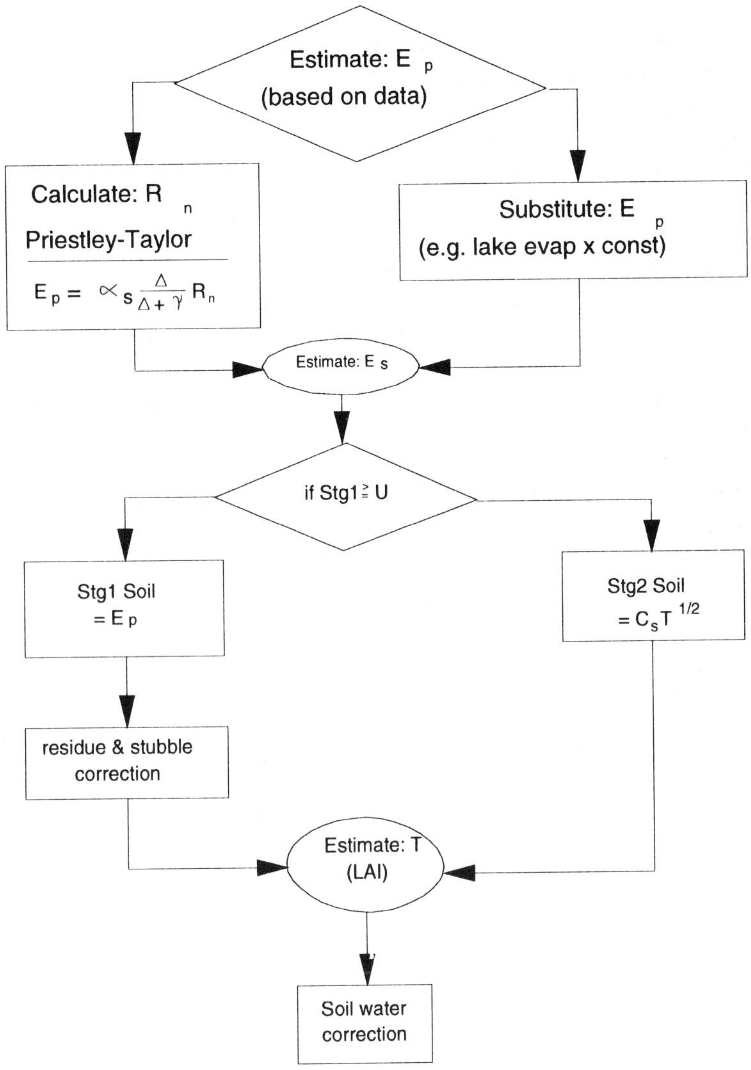

Figure 5.9
Flow chart component method

Stages of soil evaporation

Evaporation from the soil surface is known to occur in many stages (Jensen et al., 1990). Bond and Willis (1970) specifically identify three. The first is characterized by a wet soil surface and is usually known as the *constant* stage because E_s occurs at the rate of the climatic potential. The second stage is known as the *falling* stage because E_s is related to the square root of time since the stage began and is largely controlled by soil characteristics. The third stage *E* is characterized by very slow rates from a very dry soil profile. The soil surface is said to control the rate of E. Discussion of stages 1 and 2 follow.

Stage 1 evaporation

First stage soil evaporation occurs when the soil surface is wet following an irrigation or significant rain. The definition of a significant rain is not clear. When the soil surface is wet, E_s is limited by the atmospheric demand and proceeds at or near the potential rate.

Minor differences are found in references describing stage 1 soil evaporation. These differences appear to be small and should not severely impact results. E_s during stage 1 is controlled by atmospheric demand. Therefore, there is a need to calculate E_p by a suitable means. The next step is to estimate the duration of stage 1.

There is a need to distinguish among R_n, R_{no}, and R_{ns}. The nomenclature of Ritchie (1972) is followed. R_n is net radiation described in general terms as shown in Chapter 2. Ritchie specifically defines R_{no} as net radiation above any canopy and specifically refers to estimations of E_s as a separate component. R_{ns} is net radiation at the soil surface and is below any canopy.

R_n was calculated using a method credited to Ritchie and Burnett (1971). It is difficult to determine the method of calculating R_n recommended in Jensen et al. (1990). Nieber and Patwardhan (1989) do not justify the method of calculating R_n used by them. Logic indicates that the method used to calculate R_n should be appropriate for a soil surface. This means that the albedo should be appropriate for soil and should recognize any snow or ice on the soil surface.

Obviously, if no canopy exists, LAI can be considered to be zero and

$$R_{ns} = R_n \tag{5.29}$$

Ritchie (1972) at first estimated E_p by a version of the Penman equation. Usually E_p is estimated by a Priestley-Taylor equation which follows.

$$E_p = \alpha_s R_{ns} \frac{\Delta}{\Delta + \gamma} \quad (5.30)$$

where α_s is a constant used as 1.28 by Nieber and Patwardhan (1989) and is calculated by regression equations, $\alpha_s = 0.92 + 0.4(R_{ns}/R_n)$, and $\alpha_s = 1$ for LAI > 2.7 (Jensen et al., 1990).

$$E_s = 0.9\, E_p \quad (5.31)$$

where E_s is stage 1 soil evaporation in mm day^{-1}. Other references use the constant as unity in place of 0.9 (Nieber and Patwardhan, 1989).

There are many minor differences in the calculation of R_n and E_p. These differences probably do not make significant differences in estimates of soil evaporation. Careful and complete justification of all equations used is required to determine E and T by the component method.

Stage 1 soil evaporation continues until the soil surface appears to be dry. This is said to be reached when the cumulative stage 1 evaporation reaches a total amount, U mm. U is related to soil type.

Stage 2 evaporation

Stage 2 soil evaporation is characterized by a dry soil surface and the hydraulic properties of the soil control evaporation. The cumulative stage 1 soil evaporation when stage 2 begins is needed to determine when stage 2 calculations begin.

$$E_{stg2} = C_s(\sqrt{t} - \sqrt{t-1}) \quad (5.32)$$

where E_{stg2} is daily soil evaporation for stage 2 in mm, t is the time in days since stage 2 soil evaporation began, and C_s is a constant which depends upon soil type and ultimately hydraulic conductivity. C_s has the units of mm day$^{-1/2}$.

Values of U and C_s appear below. The values of U were calculated from relationships and data appearing in Nieber and Patwardhan (1989). The values shown are similar to those found in Ritchie (1972) and Jensen et al. (1990).

Table 5.23
Typical values of the constants
U in mm and C_s in mm day^{-1}

texture	C_s	U	references	year
clay loam	5.08	12.0	Ritchie, van Bavel	n.a.
Yolo loam	4.04	9.0	Ritchie, La Rue	1968
black clay	3.50	6.0	Ritchie	1972
sand	3.34	6.0	Ritchie, Black et al	1969
loam	4.50	10.6	Nieber & Patw.	1988
clay	5.50	13.2	Nieber & Patw.	1988
sand	3.30	5.4	Nieber & Patw.	1988
clay loam	5.60	13.7	Steiner disk	1989
clay loam	8.20	9.9	Steiner sweep	1989
clay loam	7.80	9.9	Steiner no-till	1989

Water needed to repeat stage 1 cycle

The amount of precipitation required to enter the soil that is needed to re-start stage 1 evaporation, when in stage 2 evaporation, is not well-defined. Ritchie (1972) indicates that for the best accuracy the stage 1 E_s can be said to occur when the top 500 mm of the soil profile returns to the DUL (field capacity). Pochop et al. (1985) used the criteria that when the top 150 mm of the soil profile reached the DUL (field capacity), stage 1 evaporation could be assumed to resume for a dry rangeland site in Colorado. One obvious assumption is that precipitation that is equal to or greater than U will trigger a repeat cycle. Precipitation amounts less than the amount required to trigger a repeat cycle will evaporate at the potential rate because even small amounts of rainfall will cause the surface soil to be wet. Ritchie (1985) says that as long as reasonable assumptions are made, good results can be obtained using the E and T component computation method.

5.8.2 *Inclusion of growing vegetation*

When plants become green and begin to grow, the soil surface is increasingly shaded, and R_{no} is reduced below R_n. R_{ns} is related to R_n by the following (Ritchie, 1972; Jensen et al., 1990).

$$R_{ns} = R_n \exp(-0.4\, LAI)\ for\ LAI < 2.7 \tag{5.33}$$

where LAI is leaf area index which is dimensionless, R_n is net radiation above the surface, and R_{ns} is net radiation at the soil surface in MJ m_{-2} day_{-1}. The constant 0.4 appears to be the constant 0.398 of Ritchie (1972) rounded up because of the precision of the method.

Leaf area index, LAI

LAI is an important parameter in fields such as engineering, but it is even more important in plant sciences. LAI is defined as *the area of one side of leaves per unit of soil surface* (Jensen et al., 1990). Hay and Walker, (1989) define LAI similarly as *the area of leaf lamina (one surface only) per unit of soil surface area.* They indicate that LAI is related to temperature, nitrogen status, population density, and water supply. This definition seems to assume green growing leaves. Some plant models utilize a parameter known as the *green leaf area index*, GAI (Hay and Walker, 1989).

No mention is made of the physiological consequences of the maturation of plants in the definition. Aging or senescence of leaves as to the expansion and decline of leaf area is poorly documented, according to Hay and Walker (1989). A small decline in LAI as plants mature because of dropping of lower leaves can be expected. The leaf area index of a maturing leaf undergoing a color change should not become reduced, but ET_{max} is known to decline as crops mature. In the previously discussed approach where combined ET is discussed, this is accounted for by means of declining values of K_c. See Section 5.4 for more information.

Hay and Walker (1989) show LAI curves as a function of time for a variety of crops and cultural variations such as nitrogen effects and planting dates. They show a decline in LAI as many kinds of crops mature with some crops such as cereals going to zero. Tables 5.24 and 5.25 show values of LAI for various crops (Environmental Protection Agency, 1984). Significant declines of from 3 to 0.5 are shown for grasses. Some crops, such as corn, even are shown with zero values of LAI at the end of the growing season.

Some of the declines seem to be harsh. A zero LAI seems to be associated with harvest and perhaps a complete removal of the crop. The LAI decreases as plants mature are possibly due in part to color changes. The LAI figures shown in Tables 5.24 and 5.25 may be viewed as broad averages with the likelihood that values may be expected to change with cultural practices. Perhaps, the severe decline at end of the growing season can be explained by contrasting the definition of the LAI with that of the GAI.

The declining values of LAI shown in Tables 5.24 and 5.25 seem severe, but similar values are shown by Hay and Walker (1989). Delaney (personal communication) indicates he has measured similar declining values of LAI in alfalfa.

Table 5.24
LAI & growing season
for fractions of the growing season
EPA (1984)

growing season	excellent grass	good row crop
0.0	0.00	0.00
0.1	1.84	0.15
0.2	3.00	0.40
0.3	3.00	2.18
0.4	3.00	2.97
0.5	3.00	3.00
0.6	3.00	2.96
0.7	2.70	2.92
0.8	1.96	2.30
0.9	0.96	1.15
1.0	0.50	0.50

Table 5.25
Table showing LAI and growing season
for specific crops
for fractions of the growing season
EPA (1984)

season	corn	oats	wheat	grass	soybeans
0.0	0.00	0.00	0.00	0.00	0.00
0.1	0.09	0.42	0.47	1.84	0.15
0.2	0.19	0.84	0.90	3.00	0.40
0.3	0.23	0.90	0.90	3.00	2.18
0.4	0.49	0.90	0.90	3.00	2.97
0.5	1.16	0.98	0.90	3.00	3.00
0.6	2.97	2.62	1.62	3.00	2.96
0.7	3.00	3.00	3.00	2.70	2.92
0.8	2.72	3.00	3.00	1.96	2.30
0.9	1.83	3.00	0.96	0.96	1.15
1.0	0.00	0.00	0.00	0.50	0.50

5.8.3 *Transpiration*

The LAI increases as plants become established and grow. As crops grow, it is logical to assume that *transpiration*, T, increases until the plants mature. Tables 5.24 and 5.25 apparently include the effects of maturity because LAI is shown to decrease toward the end of the growing season. Ritchie (1972, 1974) presents empirical relationships that can be used to estimate T as a function of LAI.

$$T = E_p[-0.21 + 0.70\, LAI^{0.5}]\,, 0.1 \le LAI \le 2.7 \qquad (5.34)$$

where E_o is potential evaporation above any canopy in mm day^{-1}, and LAI is the leaf area index which is dimensionless.

$$T = E_p\,,\ LAI > 2.7\,, PESW_{act} > 0.25 \qquad (5.35)$$

where $PESW_{act}$ is calculated by Equation 5.4.
As the soil dries, T is reduced and is equal to the following

$$T = 4\, E_p\, PESW_{act}\,, PESW_{act} < 0.25\,, LAI > 2.7 \qquad (5.36)$$

Others use a simpler formulation for the computation of T (Environmental Protection Agency, 1984; Nieber and Patwardhan, 1989)

$$T = \frac{E_o\, LAI}{3}\,, LAI \leq 3 \qquad (5.37)$$

5.8.4 *Residues*

There is a large body of references dealing with the effects of residues on soil evaporation. Residue is defined here as no stubble exists at the surface of the ground. Therefore, for example, residue may be straw from small grain.

The following apply only to stage 1 soil E. In soil evaporation modelling, no recognition of residues and/or stubble is made during stage 2 evaporation. The work of Bond and Willis (1970) provide a correction to stage 1 evaporation based upon the weight or mass of stubble per unit area. A much more recent reference is that of Steiner (1989). Steiner provides a correction based upon the depth of residue.

This discussion is based on a distinct division between residue and stubble. The studies described by Bond and Willis (1970) and Steiner (1989) may include stubble on the soil surface, but it is difficult to determine if soil tillage effects are included.

Bond and Willis (1970) present a curve showing the percent reduction in the rate of first stage soil evaporation based on the mass of residue in *kg* ha^{-1}. An algorithm leading to a numerical correction may then be easily determined.

Steiner (1989) shows a set of corrections to the rate of stage 1 soil E. The corrections are for both the depth of residue in *m* (from dividing Steiner's units which were $m^3\, m^{-2}$) and in the mass of residue in *g* m^{-2}. The Steiner (1989) corrections have the form of:

$$y = A + B\,(\ln \cdot z) \qquad (5.38)$$

where *y* is the relative reduction in the rate of stage 1 soil E, and *z* has the units shown.

Table 5.26
Residue regression constants
cotton, sorghum and wheat were pooled
Steiner (1989)

	units	A	B	r^2	n
cotton	g/m^2	2.07	-0.223	0.95	15
sorghum	g/m^2	1.72	-0.205	0.93	15
wheat	g/m^2	1.50	-0.200	0.96	6
cotton	m	-0.85	-0.223	0.95	15
sorghum	m	-0.83	-0.205	0.93	15
wheat	m	-0.91	-0.200	0.96	6
pooled	m	-0.99	-0.236	0.87	36

5.8.5 *Stubbles*

Stubble is defined as the non-green stalks left on the ground surface of a field after harvest, for example, of small grain. The stalks left after the harvest of grass for hay also is referred to as stubble.

The LAI is equal to zero during non-growing season periods because it is based on transpiring plants. The soil surface may be bare soil in many cases. However, stubble may be present on the surface of the field. The stubble shades the soil surface and reduces soil E. The Environmental Protection Agency (1984) uses a correction for the existence of stubble. The correction is the same as the correction for a crop with increasing LAI. The correction is based on the *winter cover factor*, WCF.

$$R_{ns} = R_n \exp(-0.4\ WCF) \tag{5.39}$$

EPA (1984) shows the following values for WCF.

Table 5.27
Winter cover factors
EPA (1984)

	Winter Cover Factor WCF
bare soil & row crop	0.0
excellent grass stand	1.8
good grass stand	1.2
fair grass stand	0.6
poor grass stand	0.3

5.8.6 *Adaptations for winter conditions*

The estimation of E has frequently been applied to winter conditions, for example, see Williams et al. (1985) and Andersson and Harding (1991). Jackson et al. (1976) indicate that the summer value of C_s should be divided by two for winter conditions. This change is apparently needed to ensure that estimated stage 2 soil evaporation is not greater than E_p. The following simple procedures cannot be assumed to be satisfactory for severe conditions. A climatic cutoff regarding severe conditions is not known.

Determination of the form of precipitation

The simplest method used to estimate the form of cold climate precipitation is simply to assume that when the average daily temperature is less than 0°C, snowfall is said to occur. This procedure has been applied to the United Kingdom and southern Sweden (Andersson and Harding, 1991).

Calculations of net radiation

Procedures for the estimation of net radiation should recognize winter conditions. Formulae used should recognize winter conditions. Williams et al. (1985) changed the albedo to 0.6 when over 5 mm of snow is present.

Determination of snow melt

Under many conditions, a simple degree day formula may be used to estimate snow melt.

$$MELT = C_{melt}T \tag{5.40}$$

where MELT is snow melt mm day^{-1}, and T is average daily air temperature in °C. Andersson and Harding 1991) show values of the degree day factor C_{melt} of 3.5 for grassland and 2.0 for forest.

Estimation of atmospheric water loss

Snow melt is usually considered to return to the atmosphere as evaporation except for surface runoff. In addition, water loss to the atmosphere should be equal to or less than *potential soil evaporation*, E_p. Some models have refreezing routines for snow melt that does not contribute to surface runoff and that exceeds E_p.

5.9 Irrigation water requirements

Irrigation water requirements, IWR, may be defined as *the addition of water over naturally occurring precipitation to provide soil water conditions for as near as possible optimum conditions for plant growth and vegetative production.* The determination of ET is central to the problem of determining irrigation water requirements. The determination of IWR may be viewed from the standpoint of irrigation scheduling, or from the viewpoint of irrigation water rights. Discussions in this chapter are directed toward commercial crops. These commercial crops usually include plants such as corn or alfalfa. These crops might include ornamental plants as well.

5.10 Components of irrigation water requirements

Water to be supplied by irrigation is primarily determined by ET. However, the need for irrigation water may arise from several other factors. All of the components of ET should be considered in the determination of IWR.

5.10.1 *The determination of evapotranspiration*

The determination of ET using climatic data is, of course, the main subject of this volume. The determination of IWR involves incorporating ET_a into a soil water balance procedure. The incorporation of ET_a into a soil water balance is usually neglected in most volumes dealing with ET. Calculation details for the estimation of ET_a are considered throughout this volume.

5.10.2 *Other components of irrigation water requirements*

Although irrigation is practiced to supply plants, particularly commercial crops, with water for ET, other water requirements may be essential parts of crop culture. These requirements may not be expressly defined as a part of crop production but may be required for crop growth. The additional components may not be recognized legally by irrigation water rights.

Leaching requirements

The maintenance and control of salt within the root zone is a requisite for the sustained production of commercial agricultural crops. The control of salts within the root zone is accomplished by the application of excess irrigation water over the amount required for the maintenance of optimum soil water for the best possible crop growth. Other references provide detailed discussions of leaching requirements.

Climate control

Irrigation water is occasionally applied to prevent frost damage to crops. As long as unfrozen water remains on crop surfaces, frost damage will not occur to crops. When freezing occurs, heat is released. Gerber and Martsolf (1979) discuss details of frost protection by sprinkling.

Sprinkling of crops for the mitigation of heat stress is also practiced. The evaporation of water from sprinkling involves cooling because of a transfer of latent heat. Chesness et al. (1979), Merva and Vandenbrink (1979), Rosenberg (1979) and Blad and Lemeur (1979) discuss the practice of heat stress alleviation in the ASAE Monograph *Modification of the Aerial Environment of Crops* (Barfield and Gerber, 1979). This is the same volume as the reference of Gerber and Martsolf (1979).

The application of irrigation water by sprinkling also may be used to modify the crop environment so that morphological plant development is altered. For example, bud development in Utah has been delayed until the frost probability is reduced (Griffin and Richardson, 1979.) The monograph by Barfield and Gerber (1979) references this paper and others dealing with the modification of crop development by the application of irrigation water.

Chemical applications

Chemicals, such as fertilizer and pesticides, are applied through irrigation systems. Carefully controlled irrigations are required because non-uniformities and other excess applications in irrigation result in non-uniform or excess chemical applications. Procedures for the application of chemicals through irrigations are discussed by The Irrigation Association, (Pair et al., 1983) and by Howell et al. (1980).

Other requirements

Burman et al. (1980) discuss the application of irrigation water for other purposes such as the germination of crops and dust suppression. Such practices are not widespread in usage.

5.11 Sources of water for crop growth

It is well known and obvious that plants adapt to the water and fertility conditions found at a site (Siemer and Rumberg, 1975). Some plants, such as rushes, are adapted to very wet soil water conditions. Commercial crops, such as corn, are adapted to moist conditions and do not tolerate long periods of saturation or low soil water conditions. Desert plants, such as cacti, are suited to very dry soil water conditions.

Plants require specific soil water conditions, but the origin of this water may be from various sources. The two major sources of this water are from natural precipitation on the site of the plants themselves or natural precipitation that is concentrated, transported, and applied as irrigation water. Irrigation water may be from underground aquifers, but the water is ultimately from natural precipitation.

The following discussion of components of plant water is arbitrary and not necessarily without contradiction.

5.12 Soil water storage of natural precipitation

Plants discussed in this volume require soil for structural integrity, stored water, a source for oxygen, and necessary nutrients. Naturally occurring precipitation at the sites of crops and irrigation water are temporarily stored in the surrounding soil. Emphasis on various components follows.

5.12.1 *Effective precipitation*

Effective precipitation may be defined in various ways (Dastane, 1974). In this volume *effective precipitation*, P_e, is defined as that portion of precipitation that satisfies the ET requirements of plants. This definition is consistent with the definitions of Burman et al. (1980, 1983) and Jensen et al. (1990) with the minor exception that the word precipitation is used instead of rainfall because snowfall may be utilized for ET by plants in high altitude areas.

The definition of P_e may be for different hydrologic purposes. A hydrologist interested in underground water resources may define P_e as the amount of precipitation reaching the aquifer. Dastane (1974) presents an extensive discussion on the differences in defining P_e.

This definition of P_e excludes precipitation water lost by surface runoff or lost to the root zone by deep percolation. P_e depends primarily upon the amount of water in the root zone, the frequency of precipitation, the form of precipitation, and the rate of precipitation. Moist soil near or above field capacity will lose water to deep percolation. Frequency of precipitation can change the intake characteristics of the soil. Precipitation in the form of snow may remain on the surface until melted. The maximum infiltration rate changes with time.

This definition of P_e also includes year-round precipitation. The references cited do not mention year-round precipitation specifically. Non-growing season precipitation, which is primarily snowfall in high altitude areas, is utilized by plants according to Hill et al. (1989), Walter et al. (1990), and Stewart and Hagan (1973).

Both Blaney and Criddle (1962) and the U. S. Soil Conservation (USDA 1970) recommend that non-growing season precipitation be deducted from irrigation water requirements. Non-growing season precipitation has been apparently neglected in most determinations of irrigation water right quantities.

Non-irrigated plants are supplied by year-round precipitation, water from shallow water tables, and by naturally occurring surface water that might flow onto a site. Both growing season and non-growing season soil water status must be considered in any hydrologic analysis for non-irrigated plants.

5.12.2 *Effective precipitation and area*

The following discussion will show the three precipitation losses according to the definition of P_e used in the previous section.

1. Water intercepted by plant surfaces and then evaporated into the atmosphere without striking the ground surface.

2. Surface runoff of water from precipitation.

3. Water that moves below the root zone by deep percolation.

The definition of P_e strictly applies to a point in a field or at most a small field. Often all precipitation striking a larger area, such as an entire irrigation district or farm, may be considered as being effective in meeting plant or crop needs.

The water evaporating directly into the atmosphere reduces the evaporative power of the air and, therefore, reduces ET downwind. The magnitude of this reduction has not been quantified but is nevertheless real. Burman et al. (1975) show how water in the air increases across a large irrigated area in Idaho, and at the same time potential irrigation demand decreases. Only a very detailed hydrologic analysis would show whether interception really represents a precipitation loss.

Precipitation water lost to a point certainly represents a loss of water to that point. The disposition of surface runoff water shows that surface runoff may or may not be a loss. Surface runoff may enter ditches and then in turn be applied to the land by irrigation. Surface runoff may return to the stream system and then again be diverted for irrigation. Surface runoff may be considered to be loss if it later becomes ET for undesirable plants. The definition of undesirable plants is difficult when considering wetland areas associated with irrigation.

An irrigator may assume that water consumed by ET from non-commercial plants, such as willows, represents a loss while at the same time the production of willows and similar plants may be regarded as being worthwhile by a biologist interested in wildlife. There is a tendency for recognition of these uses of water in the United States.

Water carried below the root zone by deep percolation is similar to surface runoff. Water going below the root zone often contributes water to the aquifer which may or may not be pumped for irrigation. Percolation water may become ET for plants which may or may not be considered to be undesirable plants, depending on the point of view.

5.12.3 *Effective precipitation and specific crops*

If P_e is defined as being *that portion of precipitation that satisfies the ET requirements of plants*, it may be then assumed that the definition is plant specific. If the growing season is a limiting factor, P_e would be different for a long season crop, such as sugar beets, as contrasted to a short season crop, such as small grains. Interception may not be the same for a sparse crop as contrasted to a lush crop. Periodic cuttings made of alfalfa could be expected to change interception. Deep percolation may be assumed to be plant specific because of different root zone depths and differing soil water profiles found under different crops or plant communities.

The fact that P_e is plant or crop specific is rarely discussed. The empirical methods of estimating P_e that follow do not distinguish between different crops.

5.12.4 *Growing season effective precipitation*

In many applications, part of ET is assumed to be supplied by effective growing season precipitation. Methods of estimating P_e range from very simple empirical models of questionable accuracy to complex hydrologic balance procedures that may be difficult to apply.

US Bureau of Reclamation model

A very simple method of estimating P_e which has been widely used in the US is known as the US Bureau of Reclamation method (Stamm, 1967). The method is recommended for the five driest consecutive years of record at a location based on annual precipitation. The driest years' limitation has been neglected in many cases. The method permits the estimation of monthly P_e based on a simple series of percentages. The simplicity of the method has been very attractive. The accuracy of such a simple method to a complex process is questionable.

Table 5.28
US Bureau of Reclamation method
monthly effective precipitation

Effective rainfall %	Increment of rainfall mm
95	0.0 - 25.4
90	25.4 - 50.8
82	50.8 - 76.2
65	76.2 - 101.6
45	101.6 - 127
25	127 -152.4
0	over 152.4

US Soil Conservation Service model

The analysis of a data set for 22 locations for a 50-year period produced a method known as the US Soil Conservation Service method (USDA 1970). A daily water balance was made.

Results of the method are usually presented in tabular form with P_e being a function of monthly ET, total monthly rainfall, and net depth of water depletion. The original reference utilized English units. Expressions for estimating P_e in SI Units are presented by Jensen et al. (1990).

$$P_e = f(D)\,[1.25R_t^{0.824} - 2.93]\,[10^{0.000955\,\mathrm{ET}}] \quad (5.41)$$

$$f(D) = 0.53 + 0.0116D - 8.94 \cdot 10^{-5}D^2 + 2.323 \cdot 10^{-7}D^3 \quad (5.42)$$

where f(D) is a correction factor for soil water depletion, D, and is equal to unity when D is equal to 75 mm, and R_t is monthly average rainfall in mm. ET is monthly evapotranspiration in mm. The following tables were calculated using the above two equations.

Table 5.29
Monthly effective rainfall, SCS-TR21 method
P_e, rainfall, and ET are mm

	mean monthly evapotranspiration															
ET	12.5	25.0	37.5	50.0	62.5	75.0	87.5	100.0	112.5	125.0	137.5	150.0	162.5	175.0	187.5	200.0
	mean monthly P_e for various values of mean monthly rainfall															
25	7	16	23	30												
50	8	17	24	32	39	46	52									
75	8	17	26	34	41	48	55	62	69	75						
100	9	18	27	35	43	51	58	66	73	80	86	93	100			
125	9	19	29	37	46	54	62	69	77	84	91	98	105	112	119	126
150	10	21	30	40	48	57	65	73	81	89	96	104	111	119	126	133
175	10	22	32	42	51	60	69	77	86	94	102	110	118	125	133	140
200	11	23	34	44	54	64	73	82	91	99	108	116	124	132	140	148
225	12	24	36	47	57	67	77	86	96	105	114	123	131	140	148	157

Table 5.30
f(D) correction factors for depletions
other than 75mm
SCS TR-21, soil water depletion
values are in mm
f(D) is the correction factor

Depletion	20	45	70	95	120	145	170	195
f(D)	0.73	0.89	0.98	1.02	1.04	1.04	1.06	1.11

The locations of data stations used in calibrating the method appear in Nieber and Patwardhan (1989). The method can be expected to provide questionable estimates of effective rainfall in locations radically different from the climate used for calibration.

Water balance model

There is no unique *water balance model*, WBM. WBM's may be defined as any model which evaluates the hydrologic equation term by term. It is possible to make a simple hydrologic balance using hand calculations, but it is obvious that almost all hydrologic balance calculations are made using personal, mini-, or mainframe computers.

Many hydrologic or plant growth models may be used for estimating P_e where soil properties, plant attributes, and other appropriate parameters are considered. Examples are the SPAW (Saxton et al., 1974) and the CREAMS (Knisel, 1980) models.

Patwardhan et al. (1990) assess the accuracy of a WBM especially constructed for comparative purposes to compare WBM results with those of the empirical simplifications such as the SCS method. Their model was constructed using portions of other applicable soil water balance models, and specific details are not shown. They reported that the Soil Conservation Service (USDA, 1970) and Hershfield models (1964) provided reasonable estimates for well-drained soils but not for poorly drained soils, when the SWB model was considered as standard.

5.13 Non-growing season precipitation

Previous discussion indicates that naturally occurring vegetation survives because of naturally occurring precipitation which is stored in the soil profile. Non-irrigated vegetation may also survive because of water contributions from a shallow water table and surface water that may flow to the site. Blaney and Criddle (1962), Stewart and Hagan (1973) and the Soil Conservation Service (USDA, 1970) indicate that non-growing season stored soil water is clearly a part of IWR. Blaney and Criddle (1962) used assumed soil water values in their publication, and the Soil Conservation Service (USDA 1970) recommended that winter soil water storage be evaluated in the determination of IWR amounts.

Howe and Rhoades (1962) and Howe (1951) make the point that measurements made near Scottsbluff, Nebraska, show that corn receives about 175 mm of water from stored soil water, and that beans receive about 45 mm from stored soil water.

Hill et al. (1989), in summarizing a five-year field study in the Bear River area of Utah, Wyoming, and Idaho, recommended that carry-over soil water storage be estimated as 67% of non-growing season precipitation, provided it does not exceed 75% of available soil water storage within the root zone. This is PESW, *plant extractable soil water*, which is discussed in detail in Chapter 3 of this volume.

Walter et al. (1990), recommend that the non-growing season soil water build-up be estimated as the following factors F_{wsm} to be multiplied by the non-growing season precipitation. The work cited was performed at the unusually high elevation of about 3000 m in a mountain valley in central Colorado and may be very site specific.

Table 5.45
Factors for estimating winter soil water build-up from non-growing season precipitation for variable water table conditions

F_{wsm}	Early Season Water Table
0.22	0 to 0.6 m
0.50	0.6 to 0.91 m
0.63	0.91 to 1.22 m
0.69	over 0.6 m
0.72	over 0.91 m
0.79	over 1.22 m

5.13.1 *Irrigation water from shallow water tables*

Most irrigated agriculture is located in areas where water tables are sufficiently below the root zone so that water transmitted upward from a water table may be considered to be negligible. In some cases, the location of a water table below the surface is controlled so that the water table control may be considered to be a form of irrigation in itself. In still many other cases, shallow water tables may be naturally occurring or the result of excess irrigation water combined with poor drainage. The following material deals with contributions of water from a relatively shallow water table.

Alfalfa at Grand Junction, CO, (Kruse and Neale, 1985) obtained about 91% of its water requirement from a water table located at 0.6 meters below the surface. The same authors report that corn received 66% of its water from a water table at 0.6 meters and about 37% of its water supply when the water table was located at 1.05 meters. Walter et al. (1990) discuss the origin of water for rangeland which has both irrigated grasses and grass-like plants, known as rushes. The rangeland was formerly irrigated mountain meadows and received no irrigation water. Walter et al. (1990) report that the following percentages of Et_a came from the water table: 43% with a 0.61 meter water table, 21% with a 1.22 meter water table, and about 15% for a 1.83 meter water table.

Chapter 6

PRODUCTION, VEGETATION & ET

Theoretically based equations relating biological production directly to ET are unknown to the authors. Any empirical formula relating vegetative production to ET or irrigation water applied may be defined as a *crop water production function*, CWPF. A complete coverage of CWPF information would involve a lengthy volume in itself. In addition, considerable research and developmental activity are currently underway. Simplifying assumptions are, therefore, needed to limit the scope of material covered on CWPF's in this volume.

Excellent reviews of the topic of relationships between vegetative production and ET_a include those of Vaux and Pruitt (1983), Howell (1990), and Howell et al. (1990). These reviews include additional topics such as salinity and other agronomic influences.

6.1 Uses of production functions

Production functions have the following uses as a minimum:

1. To relate vegetative, fruit, or seed production to:
   a. transpiration
   b. evapotranspiration
   c. irrigation water applications
   (perhaps the prediction of ET from production is the most important)

2. To provide design information for irrigation systems.

3. To provide water management information.

4. To account for the influence of cultural practices such as fertility on vegetative production and ET_a.

5. To make economic analyses of water management alternatives.

6.2 Defining assumptions

The following assumptions were used in preparation of the subsequent material. Many additional topics are discussed, but not in depth.

6.2.1 *Seasonal assumptions*

Plant vegetative production as influenced by soil water status is known to be highly related to the segment of the growing season under consideration. In the 1955 Yearbook of Agriculture, *Water*, Rhoades and Nelson, (1955) indicate that the growth of corn is highly influenced by drouth during the tasseling and silking portion of the growing season, while drouth during the maturation stage has only a minimal reduction on the production of grain.

A classic set of experiments relating plant growth and vegetative production to growth stage, again with the commercial crop corn, was reported by Stewart et al. (1974). The determination of CWPF values was stressed, and the influence of drouth timing with respect to growth stage was investigated. The earlier results of Rhoades and Nelson were confirmed.

More recent studies of grain sorghum response to soil water level in Nebraska were reported by Garrity et al. (1982a,b). CWPF values were sensitive to the portion of the growing season when drouth occurred and even to individual varieties.

While plant response to drouth is known to be dependent on the portion of the growing season under consideration, only those CWPF's dealing with seasonal estimates of ET are considered in detail. This is only because of the need for simplification and space limitations.

6.2.2 *Functional limitations*

Only CWPF's dealing with ET are covered in detail. Other factors such as economic analyses or cultural practices are not discussed.

6.3 Common models

Many mathematical formulae capable of generating values of ET as a function of parameters, such as relative evapotranspiration, have been fitted to measurements. Vaux and Pruitt (1983) discuss many of the formulations. The many density functions used in statistical analyses illustrate the use of mathematical formulae to match data.

6.3.1 *Linear relationships*

Many references show the results of simple linear regression and correlation between ET and vegetation production in weight, W, in kN or mass units, M in kg. These are applied to commercial crops such as alfalfa. Linear relationships will be applied because the analysis is the only one that applies to plants not classified as traditional commercial crops such as corn, alfalfa, or sugar beets. The maximum yields and maximum ET for some trees and shrubs, for example, are not logical in the same sense of commercial crops, or even known.

Simple straight lines

The simplest approach in relating production to transpiration, evapotranspiration, or to water applications is to use the simple least-squares regression analysis. The usual equation follows.

$$\mathrm{ET} = a + b \cdot prod \tag{6.1}$$

where *evapotranspiration*, ET, has been defined but is in mm day^{-1} here, *production*, prod, is in kg, a is the intercept of the regression line, and b is the slope of the regression line.

Straight lines forced through zero

Simple linear regression may be forced through zero with the result that zero water implies zero production. De Wit (1958) shows numerous examples where relationships have been forced through zero. The relationship follows.

$$\mathrm{ET} = m \cdot prod \tag{6.2}$$

where m is the slope of the regression line forced through zero.

Water use efficiency

Water use efficiency, WUE, may be defined as follows (Burman et al., 1980; Garrity et al., 1982a,b);

$$WUE = \frac{P}{\mathrm{ET}} \tag{6.3}$$

where *water use efficiency*, WUE, is in kg m^{-3}, P is vegetative *production* in kg and ET is now converted to m^3 of water.

WUE cannot be used directly as a CWPF because it is not possible to relate the change in vegetative production to a change in soil water or any other cultural variable. Water use efficiency, however, does provide a direct indication of the amount of vegetation that may be produced per unit volume of water imbibed by plants in evapotranspiration. It is a commonly reported statistic from many research projects. WUE is applicable to many plants that are not commercial crops. The WUE values of many studies are available and may be used for many comparisons (see Table 6.1).

An inspection of Table 6.1. shows that the WUE of improved meadows in the Laramie River area and the Green River areas of Wyoming are much higher than formerly irrigated meadows in the high area of central Colorado.

6.3.2 *Production reduction ratio*

A linear approach used by Doorenbos and Kassam (1979), Cuenca (1989), and others is shown below. The approach is based on relative production as an empirical function of relative evapotranspiration.

$$\frac{P_a}{P_m} = 1 - PRN\left[1 - \frac{\mathrm{ET}_a}{\mathrm{ET}_m}\right] \tag{6.4}$$

where P signifies *production* in kg, and the subscript m indicates maximum, and a indicates actual conditions. PRN is a constant of proportionality, or the slope of a straight line forced through zero, known as the *production reduction ratio*, for the specific equation shown above. Howell et al. (1990) define PRN as *the proportionality factor, which indicates the sensitivity of the crop yield to water deficits.*

Doorenbos and Kassam (1979) present values of PNR for the entire growing season, but at the same time they also present differing values for some commercial crops for different periods within the growing season.

Figures showing a linear relationship between relative evapotranspiration and relative production are common (CH2M-HILL, 1986) and Doorenbos and Kassam (1979). A minor point of departure is the fact that the axes are sometimes shifted from the first to the third quadrants of the coordinate axes.

6.3.3 *Measured production reduction related to relative ET*

Figure 6.1 applies to all plants, but the meaningful relationships exist primarily for commercial crops. Using a linear relationship between production reduction and evapotranspiration reduction, plants may be arbitrarily divided into classes related to drouth susceptibility.

Table 6.1
Selected values of water use efficiency, WUE,kg m^{-3}

Location	Vegetation	WUE	Reference	Remarks
South Park, CO	formerly irrig.	0.156	Walter et al, 1990	dry site deep water table
South Park, CO	formerly irrig.	0.272	Walter et al, 1990	1.2m water table 24% irr grass
South Park, CO	formerly irrig.	0.287	Walter et al, 1990	.7m water table 34% sedge 25% rush
Laramie, WY	impr meadow	0.306	Burman and Borrelli, 1984	.6m water table, timothy & forbs
South Park, CO	formerly irrig.	0.315	Walter et al, 1990	1.8m water table 20% rush + irr grass
India	wheat	0.324	Sharma et al, 1990	unirrigated grain yield
South Park, CO	formerly irrig.	0.333	Walter et al, 1990	0 water table 42% rush & 25% sedge
Laramie, WY	dry rangeland	0.429	Burman & Borrelli, 1984	western wheatgrass & blue gramma
Australia	cotton drip irr	0.555	Constable et al, 1990	drip & furrow irr., cracking clay
Laramie, WY	disturbed site	0.602	Burman & Borrelli, 1984	.1m water table, grasses and forbs
Merna, WY	mtn meadow	0.635	Pochop & Burman, 1987	timothy 30%
South Park, CO	formerly irrig.	0.636	Walter et al, 1990	0 water table moist site
India	rape (oil seed)	0.677	Raja & Bishnoi, 1990	average of varying irr schedules
South Park, CO	formerly irrig.	0.689	Walter et al, 1990	1m var water table irr grass
India	wheat	0.691	Sharma et al, 1990	irrigated grain yield
Seedskadee, WY	alfalfa	0.706	Pochop & Burman, 1987	second cutting
Australia	cotton furrow	0.740	Constable et al, 1990	drip & furrow irr., cracking clay
Laramie, WY	impr meadow	0.750	Burman & Borrelli, 1984	.4 m water table, high management
Farson, WY	alfalfa	0.803	Pochop & Burman, 1987	second cutting
Merna, WY	mtn meadow	0.830	Pochop & Burman, 1987	
Laramie, WY	impr meadow	0.864	Burman & Borrelli, 1984	.9m water table, mostly grazed
Horse Cr. WY	impr meadow	0.988	Pochop & Burman, 1987	
Horse Cr. WY	impr meadow	1.024	Pochop & Burman, 1987	Kentucky bluegrass 25%
Daniel, WY	alfalfa	1.103	Pochop & Burman, 1987	first cutting
Western NE	corn, dryland	1.180	Schneck. et al 1991	rotations including wheat and corn
Horse Creek, WY	mtn meadow	1.183	Pochop & Burman, 1987	sweet clover 50%
Horse Creek, WY	impr meadow	1.183	Pochop & Burman, 1987	
Northern Israel	cotton drip irr	1.201	Mahrer & Rytwo, 1991	daily drip irri, eddy correlation ET
Western NE	corn, limit irr	1.286	Schneck. et al 1991	rotations including wheat and corn
Farson, WY	alfalfa	1.297	Pochop & Burman, 1987	first cutting
Daniel, WY	mtn meadow	1.372	Pochop & Burman, 1987	
Seedskadee, WY	alfalfa	1.460	Pochop & Burman, 1987	first cutting
Daniel, WY	mtn meadow	1.782	Pochop & Burman, 1987	
Western NE	corn, full irr	1.875	Schneck. et al 1991	rotations including wheat and corn
Laramie, WY	impr meadow	2.051	Burman & Borrelli, 1984	.3 m water table, improved species
Southeast TX	sorghum	2.500	Szeicz et al, 1973	drouth effects & stomatal resistance

Plants may logically be classified as being susceptible to drouth when the production reduction exceeds the evapotranspiration reduction. Conversely, when the reverse occurs, plants may be said to be at least somewhat drouth tolerant. Similar arbitrary classifications appear in Doorenbos and Kassam (1979).

Figure 6.1 shows the relative production reduction plotted as a function of the relative evapotranspiration or deficit. Table 6.2 describes the classifications as to drouth resistance. Table 6.3 shows selected values of the slope PRN of the relative production and relative ET relationship.

6.4 Transferability issues

A major issue of practical significance is the problem of applying production functions in different geographical areas. Howell et al. (1990) indicate that irrigation method and management can directly influence the apportionment of E and T. Therefore, they state that the *constant of proportionality*, PRN, Equation 6.4 can be expected to change. Geographical and cultural changes in PRN are an important consideration

6.5 Sequence of calculations

The order of calculations assumes that the desired goal of calculations is to start with a defined, measured, or predicted order of soil water within the root zone of the plants. The final goal is to estimate the seasonal ET_a from vegetative production which may be vegetation, seed, or another indication of harvested vegetative material.

The following order of calculations assumes that cultural practices are based on regional recommended practices. Variables associated with cultural practice include the following:

1. Crop variety
2. Row spacing and plant population
3. Fertilizer applications
4. Weed control
5. Insect management

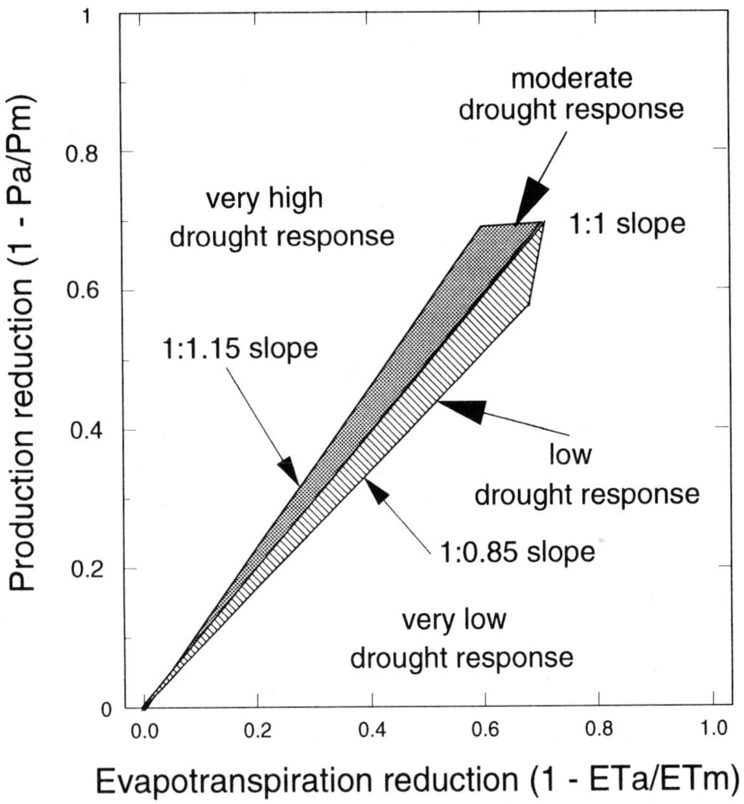

Note: lines do not approach reductions of 1 because relationships do not hold.

Figure 6.1
Relationship between relative production and relative evapotranspiration

Table 6.2
Classification of drouth tolerance

Slope	Description	Zone
>1.15	Very High Response to Drouth	VHRD
1 to 1.15	Intermediate Relative Production Response to Drouth	IRD
0.85 to 1	Low Relative Production Response to Drouth	LRD
0 to 0.85	Very Low Relative Production Response to Drouth	VLRD

Table 6.3
Measured values of the production-evapotranspiration reduction PRN

Crop	Reference	Location	PRN	r-sqr	Comments
Sorghum	Cuenca, 1989		0.98		
Sorghum	Garrity et al. 1982	Nebraska, USA	1.11	0.98	Rs 626trt GGG
Sorghum	Garrity et al. 1982	Nebraska, USA	1.47	0.93	Rs 626trt GII
Beans, Pink	Cuenca, 1989		1.50		
Sorghum	Garrity et al. 1982	Nebraska, USA	1.62	0.90	Rs 626trt GGI
Beans, Pinto	Cuenca, 1989		2.33		
Sorghum	Garrity et al. 1982	Nebraska, USA	2.43	0.95	Rs 626trt IIG

The following discussion assumes that the only variable in the determination of production is the progression of soil water. Of course, the variables listed above are important in the determination of maximum production and, hence, maximum evapotranspiration.

6.5.1 *Selection of crop water production functions*

The consulting firm CH2M HILL (1986) suggest the four possibilities to be followed in the selection of CWPF formulae. The following itemization has been adapted from CH2M HILL with the alternative of adjusting generalized CWPF's being added.

1. Select CWPF's developed from local research.
2. Select regional average CWPF's.

3. Select generalized CWPF's.
4. Adjust generalized CWPF's for local conditions.
5. Develop CWPF's from local interviews.

CH2M HILL recommend that Doorenbos and Kassam (1979) be the principal source for generalized CWPF information. At this time, other sources should be considered as well. Regional CWPF's for the US are presented by CH2M HILL (1986).

6.5.2 *Maximum production*

Maximum production for good cultural practices and near-optimum soil water conditions may often be determined with sufficient accuracy by a careful review of local statistics and necessary adjustment based upon experience.

Doorenbos and Kassam (1979) present two detailed procedures for the estimation of P_m. The first involves the estimation of the production of a standard crop for cloudless conditions. Corrections for climate and air temperature, plant species, and harvested portion then follow. The second is more detailed but involves similar corrections with the addition of a correction for the leaf area index.

6.5.3 *Maximum evapotranspiration*

The only limitation to plant growth discussed in Chapters 4 and 5 is limited soil water. For this reason, the procedures in Chapters 4 and 5 may be followed using the assumption that soil water is near optimum for vegetative production.

6.5.4 *Existent evapotranspiration*

Procedures in Chapter 5 discuss procedures for adjusting reference crop ET for plant species and soil water limitations. ET_a for limited soil water may be determined by the following methodology.

1. Measured soil water conditions.
   a. Using neutron soil water meter.
   b. Using gravimetric sampling.
   c. Estimation by probing and *feeling* samples.
2. Assumed sequence of soil water magnitudes.
3. Use of a computerized soil water model.

6.6 Flow chart for estimating vegetative production

The following flow chart has been prepared to provide a graphical illustration of sequence of calculations involved in the use of crop water production functions, CWPF's. Because this volume concerns the prediction of ET from climatic data, this chart assumes that ET may be predicted from production data.

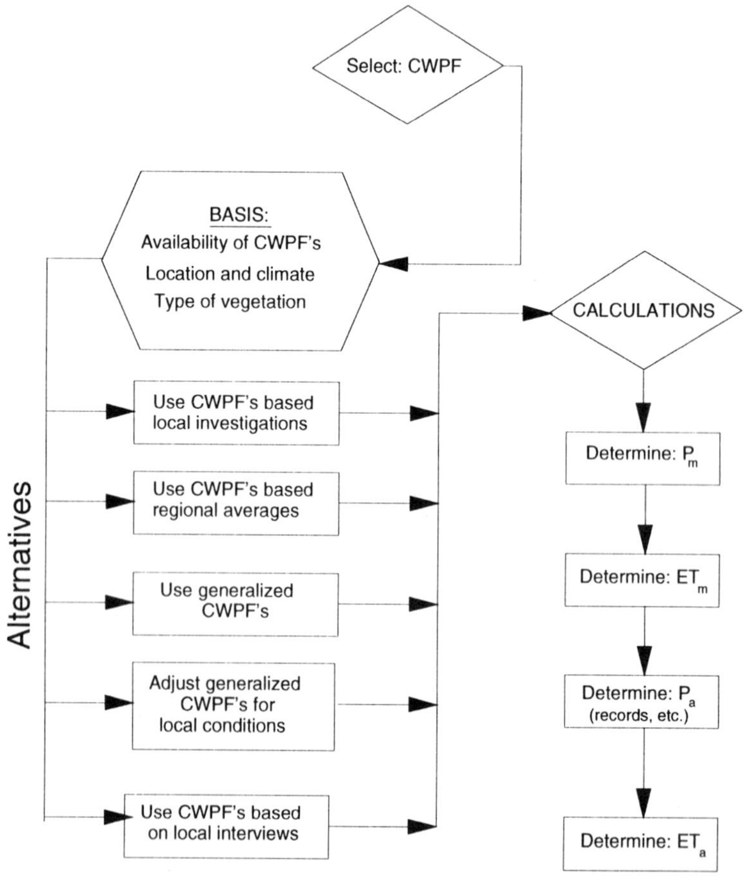

Figure 6.2
Flow chart for crop water production functions

Chapter 7 Evaporation from water surfaces

EVAPORATION FROM WATER SURFACES

Evaporation is the conversion of liquid water to water vapor. Evaporation from a free water surface, like evapotranspiration, depends on an energy supply, a difference in vapor pressure between the surface and atmosphere, and an exchange of surface air with surrounding atmospheric air. Empirical methods for estimating evaporation usually account for these factors through inclusion of temperature, radiation, humidity, and/or wind terms.

The ability to estimate evaporation from streams, lakes, and reservoirs is important to the management of our water resources. Methods for estimating evaporation are similar to those used for estimating evapotranspiration and may be classified as:

(a) empirical methods using meteorological parameters and based, in general, upon Dalton's law.

(b) water-budget methods which depend upon estimating the water inflows and outflows.

(c) energy-budget methods which require defining all energy components contributing to the energy balance.

(d) mass transfer methods which attempt to identify the transfer rate of water vapor from the evaporating surface to the air (Gangopadhyaya, 1965, and Grando, 1973).

Only the empirical methods based on meteorological parameters will be presented herein.

There are numerous factors which must be accounted for in estimating evaporation from a free water surface using empirical methods. Among the most important are the extent of the surface area, the depth of the water, and the conditions of the surroundings. Usually, the greater the surface area, the lower the evaporation rate, because water added to the air through evaporative processes modifies the air itself. Depth tends to affect water temperature and, in particular, greater depths usually lead to annual water temperature cycles which lag the annual air temperature cycle. Evaporation rates from water surfaces will usually be greater in arid than in humid environments, especially in locations having considerable advection.

The SI system of units is followed as closely as possible throughout this document *except in this chapter*. Some of the following material has units such as the *cm* even though the centimeter is not a recommended SI unit. The following methods were calibrated using archaic units, and some of the more complex equations have units which would be difficult to convert to true SI units from a practical standpoint. Therefore, units such as *cm* have been retained.

7.1 Methods

Most of the methods which have been developed to estimate evaporation from free water services are based upon Penman's combination approach (Penman, 1948). Penman's equation described evaporation as the combination of water loss due to radiation energy input and an aerodynamic term accounting for the removal of water vapor from a saturated surface. Penman's equation requires temperature, radiation, humidity, and wind data.

Several modifications of Penman's equation have appeared. Because of the general scarcity of data for some of the metrological parameters, a few of the modifications have been proposed to permit estimation of evaporation using more commonly available data. Others have been proposed to provide what are termed improved estimates of evaporation, but generally these methods have even greater data requirements. Modifications to be presented here include: Kohler-Nordenson-Fox (1955), Kohler-Parmele (1967), Linacre (1967a), Priestley-Taylor (1972), Stewart-Rouse (1976), and deBruin (1978).

The data requirements of the various evaporation formulas are analogous to those of the various ET formulas. For example, the Linacre formula requires only temperature data; the Stewart-Rouse and Priestley-Taylor equations require temperature and radiation data; the deBruin equation requires temperature, humidity, and wind data; and the Kohler-Nordenson-Fox and Kohler-Parmele equations require temperature, humidity, wind, and radiation data. The recommended applications of these equations are generally similar to those of the ET formulas, with the equations using more data recommended for estimates as short term as daily, and those using limited data recommended for longer term estimates.

Although all the equations to be discussed have been developed to estimate free-water evaporation, their applications differ greatly. Depending on the equation, development and calibration have been for Class A pan evaporation, shallow lakes, free water surfaces, etc. Extreme care should be taken to observe the conditions to which each equation applies.

7.1.1 *Kohler-Nordenson-Fox equation*

Pan evaporation

The Kohler-Nordenson-Fox equation (Kohler et al., 1955) is perhaps the model which has been most widely used for estimating evaporation. The model was an adaptation of the Penman equation for estimating Class A pan evaporation and is given as:

$$E_p = \frac{\Delta R_n + \gamma_p E_a}{\Delta + \gamma_p} \tag{7.1}$$

where E_p is *pan evaporation* in mm day^{-1}, R_n is *net radiation* in units of equivalent depth of water in mm day^{-1}, Δ is the slope of the saturation vapor-pressure versus temperature curve at the air temperature in kPa °C^{-1}, γ_p is the *psychrometric coefficient* in kPa °C^{-1}, and E_a is an aerodynamic function in mm day^{-1}.

An adjustment was made in the psychrometric coefficient to account for the sensible heat conducted through the sides and bottom of the pan. Based on studies from Lake Hefner, OK, the psychrometric coefficient was found to be 0.001568P in units of kPa per °C where P is atmospheric pressure in kPa. Kohler-Nordenson-Fox actually recommended a constant value of 0.152 for the psychrometric coefficient up to the elevation of Lake Hefner, OK, of 1525 meters. These values exceed the theoretical values for the psychrometric coefficient as calculated using the equation given in Chapter 2.

$$\gamma_p = 0.001568\, P \tag{7.2}$$

The aerodynamic function E_a was evaluated using pan data from four locations scattered across the US.

$$E_a = 25.4[0.296\{(e_a - e_d)^{0.88}\}\,(0.37 + 0.00255 u_p)] \tag{7.3}$$

where E_a is expressed in mm day^{-1}, u_p is *wind speed* at 0.152 meter above the rim of a Class A pan in km day^{-1}, e_s is *saturation vapor pressure* at the air temperature in kPa, and e_d is the *saturation vapor pressure* at the dew point temperature in kPa.

The effective net radiation for a Class A pan accounting for the effects of sensible heat transfer through the sides and bottom of the pan was determined by rearranging the pan evaporation equation to solve for R_n (Kohler et al., 1955; Lamoreux, 1962; and Jensen, 1974).

$$R_n\Delta = 154.4\exp[(1.8T_a - 180)\,(0.1024 - 0.01066 1n(0.239R_s)) - 0.01544] \tag{7.4}$$

where T_a is *air temperature* in °C, and R_s is *solar radiation* in J cm^{-2} day^{-1}. Kohler, et al. (1955) recommended that the slope of the saturation vapor pressure versus temperature curve can be estimated as:

$$\Delta = \frac{(e_s - e_a)}{(T_a - T_{dp})} \tag{7.5}$$

where T_{dp} is *dew point temperature* in °C. However, Δ can be calculated with greater accuracy using the equation given in Chapter 2.

Lake evaporation

Kohler et al. (1955) assumed that so long as air and pan water temperatures are equal, lake evaporation is 70% of that from a Class A pan. The only other change in calculating evaporation for a lake is that the value of the psychrometric coefficient, as defined by the Bowen ratio, is used in place of the value determined for a Class A pan, which becomes

$$\gamma_l = 0.000661P \tag{7.6}$$

Thus, lake evaporation is estimated by

$$E_l = 0.7\left[\frac{(\Delta R_n + \gamma_l E_a)}{(\Delta + \gamma_l)}\right] \tag{7.7}$$

7.1.2 *Kohler-Parmele equation*

Kohler-Parmele (1967) reexamined the assumptions of the Penman (1948) and Kohler-Nordenson-Fox (1955) equations. Assumptions involving the water and air temperatures being equal, sensible heat transfer through the sides and bottom of a Class A pan, net radiation calculations, the wind function E_a, advected energy, and the slope of the saturated vapor pressure versus temperature curve were all considered to have potential to contribute to errors.

Kohler-Parmele (1967) reasoned that although the Penman equation adjusts for the difference in air and surface temperatures in the aerodynamic portion of the equation, adjustments are not made to account for this difference in the energy term. Thus, they modified the original Penman type equation to account for the difference between air and water temperatures in calculating emitted radiation from the water surface. Their expression for evaporation from free water surfaces is:

$$E = \left\{ \Delta(R_{ir} - \varepsilon\sigma T_a^4) + E_a\left[\gamma + \frac{4\varepsilon\sigma T_a^3}{f(u)}\right]\right\} / \{\Delta + \gamma + \frac{4\varepsilon\sigma T_a^3}{f(u)}\} \tag{7.8}$$

where E, *evaporation*, is expressed in mm cm^{-2} day^{-1}, Δ and γ are defined as before but are to be calculated using the equations from Chapter 2, R_{ir} is the difference between incident and reflected all-wave radiation in units of equivalent depth of evaporation in mm cm^{-2} day^{-1}, ε is the *emissivity ratio* of the water surface, σ is the Stefan-Boltzmann coefficient in units of equivalent depth of evaporation, 7.81 x 10^{-11} mm cm^{-2} K^{-4} day^{-1}, T_a is the mean *air temperature* in absolute units, K, and E_a and f(u) are related by:

$$E_a = (e_s - e_a)f(u) \tag{7.9}$$

where E_a is expressed in mm cm^{-2} day^{-1}, e_s and e_a are the saturated and actual *vapor pressures* of the air in kPa, and f(u) is:

$$f(u) = 1.361 + 0.01102u_4 \tag{7.10}$$

where f(u) is expressed in mm cm^{-2} kPa^{-1} day^{-1}, and u_4 is *wind run* at 4 meters height in km day^{-1}.

Energy storage and advection were accounted for by Kohler and Parmele (1967) in estimating lake evaporation as:

$$E_L = E + \alpha[Q_i - Q_o - (S_2 - S_1)] \tag{7.11}$$

where E_L is estimated *lake evaporation*, and E is the computed free-water evaporation from the above equation both in units of mm cm^{-2} day^{-1}, Q_i and Q_o are the energy content of inflow and outflow to the lake (same units as evaporation), S_1 and S_2 are the energy storages at the beginning and end of the period (same units as evaporation), and α is expressed as:

$$\alpha = \Delta / \left\{ \Delta + \gamma + \frac{4\varepsilon\sigma T_a^3}{f(u)} \right\} \tag{7.12}$$

The ratio α is considered to be somewhat similar to the Bowen ratio. It should be used with wind measured directly above the surface, although Kohler and Parmele (1967) recommended that wind data from the 4 meter height was satisfactory.

The major difficulty in computing lake evaporation is in obtaining data for the advection, energy storage, and water temperature terms. Kohler and Parmele (1967) state that generally the advected energy is relatively small and that the energy storage may be neglected in computing mean annual evaporation. When calculating daily evaporation, daily changes in energy storage usually cannot be determined with sufficient accuracy. Thus, the equation for generalized free-water evaporation remains the most useful as a evaporation reference. Estimating specific lake evaporation is difficult, and the effect of lake size is not accounted for anyway.

7.1.3 *Priestley-Taylor equation*

Priestley and Taylor (1972) attempted to describe evaporation using the concept of equilibrium conditions, i.e. when the air in contact with a wet surface is vapor saturated. Under these conditions, radiant energy dictates the evaporation rate, and the aerodynamic term in the Penman combination equation becomes zero with the equation reducing to the energy term. However, since actual atmospheric conditions can only approach equilibrium conditions, Priestley and Taylor introduced an empirical coefficient that is defined as the ratio of evaporation from a wet surface when conditions of minimal advection exist to evaporation under equilibrium conditions. The Priestley-Taylor equation became:

$$E = \alpha_1 \left[\frac{\Delta}{\Delta + \gamma} \right] (R_n + G) \tag{7.13}$$

where the *evaporation*, E, from a uniformly saturated surface is expressed as mm cm^{-2} day^{-1}, and is defined as before, Δ and γ are calculated from the equations of Chapter 2, R_n is *net radiation* (e.g. as defined in the Kohler-Nordenson-Fox equation), G is *subsurface heat* flow in mm cm^{-2} day^{-1}, and α_1 is the empirical coefficient and is dimensionless. Using several sets of data from various surfaces having unlimited water supplies, Priestley and Taylor (1976) found an average of $\alpha_1 = 1.26$.

The difficulty with the Priestley-Taylor equation is the lack of data for R_n and G, especially G. Some users have elected to ignore G and calculate evaporation using estimated net radiation. If this is the case, then only temperature and radiation data are required. Estimates of lake evaporation using the Priestley-Taylor equation require inclusion of the G term. The Priestley-Taylor equation has been applied to ET as well.

7.1.4 *Stewart-Rouse equation*

Stewart and Rouse (1976) modified the Priestley-Taylor formula through elimination of the subsurface heat flow term and substitution of solar radiation for net radiation. The subsurface heat flow term was eliminated by considering periods of two weeks to a month and finding that the term approaches zero for these time periods. Net radiation was determined by developing a linear relationship between net radiation and solar radiation. Solar radiation may be either measured or evaluated from cloudiness or percent sunshine. The modified expression is:

$$E_{SL} = 1.26\left[\frac{\Delta}{\Delta+\gamma}\right](0.3774 + 0.7340R_{se}) \tag{7.14}$$

which reduces to

$$E_{SL} = \left[\frac{\Delta}{\Delta+\gamma}\right](0.4755 + 0.9248R_{se}) \tag{7.15}$$

where E_{SL} is the rate of *evaporation* from a shallow lake or pond expressed in mm cm^{-2} day^{-1}, Δ and γ are defined as before, and R_{Se} is the solar radiation as the equivalent depth of evaporation in mm cm^{-2} day^{-1}.

Stewart-Rouse tested the equation for summer months on shallow lakes ranging from depths of 0.5 to 2 meters. Evaporation was estimated to within 10% when periods of two weeks to a month were evaluated.

7.1.5 *deBruin equation*

deBruin (1978) derived a model to estimate shallow lake evaporation by combining the Penman and Priestley-Taylor equations. He eliminated the net radiation R_n and subsurface heat flow G terms by solving the Priestley-Taylor equation for the difference R_n-G and substituting this into the Penman equation. The result is:

$$E = \left[\frac{\alpha_1}{\alpha_1 - 1}\right]\left[\frac{\Delta}{\Delta+\gamma}\right]f(u)(e_s - e_a) \tag{7.16}$$

where E is *evaporation* from a free water surface expressed as mm cm^{-2} day^{-1}, Δ and γ are defined as before, e_s and e_a are the saturated and actual *vapor pressures* in kPa, α_1 is the empirical coefficient in the Priestley-Taylor equation, and f(u) is a wind function proposed by Sweers (1976):

$$f(u) = 2506 + 21.0u_2 \tag{7.17}$$

where the wind function is expressed as mm cm^{-2} kPa^{-1} day^{-1}, and u_2 is *wind speed* at 2 meter height in km day^{-1}. Shallow lake evaporation is estimated when the value of α_1 is taken as 1.26.

The model was tested for a shallow lake with an average depth of three meters in the Netherlands during July through October of 1967. A coefficient of determination of 0.94 was calculated when the model was compared with energy budget calculations. All input parameters were measured at the center of the lake, and periods of ten days or greater were used.

7.1.6 *Linacre equation*

Linacre (1967) developed an approximation of the Penman equation to reduce the climatic data input to only temperature. The derivation requires considerable effort and uses average values for such parameters as the albedo, specific heat and density of air, and the latent heat of evaporation. The resulting expression for evaporation from water is:

$$E = \frac{\frac{700T_m}{(100-A)} + 15(T_a - T_{dp})}{80 - T_a} \tag{7.18}$$

where *evaporation*, E, is expressed as mm day^{-1}, A is *latitude* in degrees, T_a is the mean *air temperature* in °C, T_{dp} is the mean *dew point temperature* in °C, and T_m is defined as:

$$T_m = T_a - 0.006h \tag{7.19}$$

where h is *elevation* in meters. Data input for this equation is reduced to temperature, humidity, elevation, and latitude.

Linacre recommended two methods for eliminating the need for humidity measurements. The quantity (T_a - T_{dp}) is estimated either from tabulated data developed using data from Australia and New Zealand, or by the following expression developed using data from Africa, Australia, and South America:

$$(T_a - T_{dp}) = 0.0023h + 0.37T + 0.53R + 0.35R_{ann} - 10.9 \tag{7.20}$$

where R is the average difference between mean daily maximum and minimum temperature, and R_{ann} is the difference between mean monthly temperatures of the hottest and coldest months.

As expected, the errors of the estimates increase as the time period decreases. Also, other methods (Pochop et al., 1973) may be available to estimate the dew point temperature more accurately for a specific location.

Chapter 8 Comparisons and example calculations

COMPARISONS AND EXAMPLE CALCULATIONS

Virtually all parameters calculated in estimating ET involve approximations or constants determined by a form of curve fitting or regression and correlation. This extends from intermediate calculations such as net radiation to the final determination of ET from plants. Thermodynamic properties such as latent heat of vaporization involve empirical estimates of property data found in standard references such as *Steam Tables* (Keenan et al., 1979). Very few calculations are made by formulae that may be fully justified by mathematically defensible theory. However, many methods and calculations are based upon formulae that may have a good theoretical basis.

Estimates often have considerable economic consequences. If possible methods used should be verified by comparisons of estimates with measurements. Practicing hydrologists often benefit from a detailed examination of example calculations. Many references that discuss equations used to estimate ET neglect details involving units. The practitioner often needs to be able to follow examples in a step by step manner.

8.1 Introduction

The use of the Penman method and later versions of combination theory such as the Monteith addition, (Penman, 1963; Monteith, 1965), require estimating factors such as vapor pressure, air density, atmospheric pressure, and latent heat of vaporization. Some of these factors also are required for some radiation methods such as the FAO radiation method (Doorenbos and Pruitt, 1977). Typically, methods of estimating meteorological parameters have evolved through practice. Inconsistencies appear in the literature because of piecemeal approaches and the need for simplification. These inconsistencies usually have minimal effect on the final estimates of ET but may be in error from the standpoint of basic principles. Future practitioners may need a more basic understanding of the correct applications of thermodynamic principles. With widely increasing use of inexpensive computers, more precise methods are easy to translate into computer algorithms. There is little need to use simplified procedures to avoid laborious calculations.

8.2 Example calculations

The use of the many formulae involved in estimating evapotranspiration invariably involves interpretations, assumptions, and the mundane problem of using proper units. The following illustrates the practical applications of various procedures. The illustrations have been checked using hand calculators and independent computer programs written by the authors. The SI System of units is used as closely as possible. Unit analysis is performed in a manner similar to that shown in common texts in engineering science (Fox and McDonald, 1985).

Climatic data for E and ET estimates are not shown as being from a specific location but rather were selected as being appropriate numerical values for the calculations being illustrated.

Most of the following expressions have the equation numbers assigned to them when each equation appeared in the body of this book. For example, (2.7) means that the expression had the number 2.7 originally. For an explanation the reader merely has to find Equation 2.7 in Chapter 2.

The order of calculations may be different than other publications. Very few users will set up and control the collection of climatic data. The user therefore, must use data that has been collected. The order of calculations that the hydrologist must follow in making estimates of ET will depend upon procedures followed by organizations that collect climatic data. For example, atmospheric humidity is commonly calculated from historic data. Some organizations record atmospheric humidity as percent relative humidity while other organizations record atmospheric humidity data in the form of average dew point temperature.

8.2.1 *Significant figures*

The modern calculator and computer may tempt the user to record numerical answers having figures that are not meaningful because of the numbers appearing in a print-out or visible on a screen. The general subject of significant figures is commonly addressed in books about engineering surveying (Brinker and Wolf, 1977). In general the number of significant figures in a number is equal to the number of certain numbers plus one additional digit. Thus a value of R_n shown as 25.4 MJ m^{-2} day^{-1} has 3 significant figures and it should be viewed as the 25 is certain and the 4 is questionable.

References show rules for determining significant figures for the result of addition or subtraction, products and quotients. The authors have found that from a practical standpoint standard practice is often followed. For example, most values of ET and radiation are shown as having 3 significant figures. Experience shows that in many cases only two significant figures are justified but custom is to show three. The authors have generally followed the convention of showing and using all figures shown in equations for intermediate calculations but rounding answers back to three significant figures.

There may be minor round-off differences between the accuracy found in personal computers and various hand calculators. In most cases these round-off errors will be in the third or fourth figures or beyond.

8.2.2 *Conversion between energy and depth units*

Many calculations considered in this volume involving ET result in the units of the energy equivalent of ET instead of depth per unit time. It is illogical and confusing to the user to have two units for ET, which is quantified as volume or depth per unit area per day, mm day^{-1}. For this reason ET always has depth per time units and those equations which result in energy units are shown as λ ET. Therefore, careful consideration of the units involved is essential. The ability to convert units between energy and volume units is of practical significance.

8.2.3 *SI unit principles*

General estimates and units

An example of the general use of variables and their associated units is the estimation of the psychrometer coefficient, γ. Units associated with variables follow the laws of algebra. The following is a general listing of the equation and units. For a specific application see Section 8.3.2.

$$\gamma = \frac{C_{pm} P}{\varepsilon \lambda} = \left[\frac{kJ}{kg \cdot K}\right]\left[\frac{kPa}{1}\right]\left[\frac{kg}{kJ}\right] = \frac{kPa}{K} \qquad (2.25)$$

where γ is the *psychrometer coefficient* in kPa K^{-1}, C_{pm} is *specific heat at constant pressure* of moist air in kJ kg^{-1} K^{-1}, P is *atmospheric pressure* of moist air in kPa and λ is *latent heat of vaporization* used here as kJ kg^{-1}. ε is the ratio of the mole weight of water to air and is dimensionless.

Depth and energy units

The relationship between MJ m^{-2} day^{-1} and mm day^{-1}, involves the following units. Standard SI definitions of units are used, plus the assumption that the density of water is nearly constant with a cubic meter of water having a mass of 1000 kg. This is only strictly true at 4 °C; however, the density of water, ρ_w, varies only about 1.2% between temperatures of 0 °C to 50 °C. It is common to assume that any resulting differences are negligible.

The unit principles follow:

$$\lambda \, \text{ET} = (energy) = \left[\frac{MJ}{kg}\right]\left[\frac{mm}{day}\right]\left[\frac{1000\,kg}{m^3}\right]\left[\frac{m}{1000\,mm}\right] = \frac{MJ}{m^2\,day}$$

The units of λ may be changed from kJ kg^{-1} to MJ kg^{-1} as follows:

$$\lambda = \left[\frac{kJ}{kg}\right]\left[\frac{M}{1000\ k}\right] = \frac{MJ}{kg}$$

The method of the conversion of energy to depth follows:

$$\text{ET} = [\frac{MJ}{m^2 day}][\frac{kg}{MJ}]\left[\frac{m^3}{1000kg}\right]\left[\frac{1000mm}{m}\right] = \frac{mm}{day}$$

where λ is *latent heat* of *vaporization* or *enthalpy* of *vaporization* in MJ kg^{-1} and the remainder are standard SI units.

On the surface it appears that energy in MJ m^{-2} day^{-1} may be converted to depth in mm day^{-1} by simply a division by λ when in reality the conversion also involves a multiplication and division by the number 1000 which of course cancels.

8.3 Chapter 2 examples

A few parameters may be assumed to be constant for a particular location. Even though nearly all parameters change with time, average values may be used for <u>some</u> calculations because estimated ET values are only changed slightly by the use of the average values.

8.3.1 *Constant parameters*

ATMOSPHERIC PRESSURE

given: an elevation of 1830 m above sea level, estimate the atmospheric pressure P.

<u>simple model</u>

$$P = 101.3 - 0.01055 \cdot z = 101.3 - 0.01055 \cdot 1830 = 82.0\ kPa \qquad (2.9)$$

<u>complex model</u>

$$pwr = \frac{g}{\alpha R} = \left[\frac{9.81\ m}{\sec^2}\right]\left[\frac{km}{6.5\ K}\right]\left[\frac{kg\ K}{0.287\ KJ}\right] = 5.259 \quad \text{dimensionless} \qquad (2.11)$$

see Section 2.2.1 for values of the variables $T_{o\,(K)}$ P_o and α.

using the S.I. definition of Newton's second law:

$N = \frac{1\ kg \cdot m}{\sec^2}$ and $J = N \cdot m$

$$pwr = \left[\frac{kg}{N}\right]\left[\frac{m}{\sec^2}\right] = \frac{N}{N} = dimensionless$$

$$P = P_o\left[\frac{T_{o(K)} - \alpha(z - z_o)}{T_{o(K)}}\right]^{pwr} = 101.3kPa\left[\frac{(288 - 6.5 \cdot 1.83)}{288}\right]^{5.259} = 81.1\ kPa$$

(2.10)

8.3.2 *Thermodynamic & atmospheric parameters*

WIND HEIGHT ADJUSTMENT

given: wind speed of 7.1 m $\sec^{-1}$ at an elevation of 10 meters above the surface. Assume a grass coverage of 90% with a height of 0.3 m. Convert the wind speed to an elevation of 2 m.

simplified model with exponent a = 0.2

$$u_2 = u_m\left[\frac{z}{z_m}\right]^a = 7.1\left[\frac{2}{10}\right]^{0.2} = 5.15\frac{m}{\sec} \quad (2.7)$$

complex model

$$d = 0.85\ F_c z_v = 0.85 \cdot 0.9 \cdot 0.3 = 0.229\ m \quad (2.4)$$

$$z_o = 0.13(z_v - d) = 0.13\ (0.3 - 0.229) = 0.00923\ m \quad (2.5)$$

$$u_2 = u_m\left\{\ln\left[\frac{z_2 - d}{z_o}\right] / \ln\left[\frac{z_m - d}{z_o}\right]\right\} = 7.1\left\{\ln\left[\frac{2 - 0.229}{0.00923}\right] / \ln\left[\frac{10 - 0.229}{0.00923}\right]\right\} = 5.36\frac{m}{\sec} \quad (2.8)$$

PSYCHROMETRIC

given: an air temperature of 18.3 °C and a relative humidity, RH, of 43% at an elevation of 1830 m. Determine e_a e^o_{dp}, ρ_a, γ, Δ, T_{dp} and T_{wb}.

Vapor pressure e

First calculate saturation vapor pressure, e^o. If the dew point temperature is known, the saturation vapor pressure at that point is the ambient vapor pressure. If the dew point temperature is not known, then the ambient vapor pressure is equal to the saturation vapor pressure times the relative humidity. To ensure correct calculations, the correct form of an empirical equation must be used along with the specified temperature units. The saturation vapor pressure corresponding to any desired temperature in °C is then estimated.

$$e^o = e^o_{18.3} = 0.6108\exp\left[\frac{17.27 \cdot T}{T+237.3}\right] = 0.6108\exp\left[\frac{17.27 \cdot 18.3}{18.3+237.3}\right] = 2.10\ kPa \qquad (2.14)$$

$$e_a = RH \cdot e^o_{db} = 0.43 \cdot 2.10 = 0.903\ kPa \qquad (2.22)$$

$$e_a = e^o_{dp} = 0.903\ kPa$$

$$e^o_{dp} = 0.903\ kPa = 0.6108\exp\left[\frac{17.27\,T_{dp}}{237.3+T_{dp}}\right] \qquad (2.14)$$

dew point temperature

The dew point temperature may be determined by the use of some spreadsheets or solver software to get T_{dp} = 5.6 °C or after rearranging (2.14)

$$T_{dp} = \frac{237.3\ln(0.903)+116.8}{16.78-0.378\cdot 0.903} = 5.6°C$$

specific humidity

$$q = 0.622\frac{e_a}{P-0.378e_a} = 0.622\left[\frac{0.903}{82.0-0.378\cdot 0.903}\right] = 0.006878 \qquad (2.18)$$

specific heat at constant pressure

$$C_{pm} = (1+0.84\,q)C_{pda} = (1+0.84\cdot 0.006878)\ 1.004 = 1.010\frac{kJ}{kg\ K} \qquad (2.23)$$

latent heat of vaporization λ

λ is calculated using an empirical formula intended to reproduce values of λ or *enthalpy* from steam tables. The only consideration involving units is to use the correct formula with its specified temperature.

$$\lambda = 2501 - 2.3601 \cdot T = 2501 \frac{kJ}{kg} - 2.3601 \frac{kJ}{kg\ °C} 18.3\ °C = \frac{2458\ kJ}{kg} \qquad (2.24)$$

psychrometer coefficient γ

$$\gamma = \frac{C_{pm} P}{\varepsilon \lambda} = \left[\frac{1.010\ kJ}{kg \cdot K} \right] \left[\frac{82.0\ kPa}{1} \right] \left[\frac{1}{0.622} \right] \left[\frac{kg}{2458\ kJ} \right] = 0.0542 \frac{kPa}{K} \qquad (2.25)$$

delta Δ

$$\Delta = \frac{2503}{(T + 237.3)^2} \exp \left[\frac{17.27 \cdot T}{(T + 237.3)} \right] \qquad (2.15)$$

$$\Delta = \frac{2503}{(18.3 + 237.3)^2} \exp \left[\frac{17.27 \cdot 18.3}{(18.3 + 237.3)} \right] = 0.1319 \frac{kPa}{K}$$

air density ρ

The ideal gas law is used to estimate air density, and, therefore, temperature must be in absolute units, K. Daily average air temperature is used in this example.

$$T_{(K)}' = \frac{T_K}{1 - \frac{0.378 \cdot e_a}{P}} = \frac{18.3 + 273.15}{1 - \frac{0.378 \cdot 0.903}{82.0}} = 292.7\ K \qquad (2.13)$$

remember that a J = $N \cdot m$ and that a Pa equals $N \cdot m^2$.

$$\rho_a = \frac{P}{R\, T_{(K)}'} = \left[\frac{82.0\ kN}{m^2} \right] \left[\frac{K\ kg}{0.287\ kNm} \right] \left[\frac{1}{292.7\ K} \right] = 0.976 \frac{kg}{m^3} \qquad (2.12)$$

wet bulb temperature, T_{wb}

$$e_{wb}^o = 0.6108 \exp \left[\frac{17.27\ T_{wb}}{237.3 + T_{wb}} \right] \qquad (2.14)$$

$$\Delta T_{wd}\gamma = (T_a - T_{wb})\gamma = e^o_{wb} - e_a \qquad (2.20, 2.21)$$

$$(18.3 - T_{wb})0.0542 = e^o_{wb} - 0.903$$

using solver software T_{wb} = 10.9°C
e^o_{wb} = 1.305kPa

RADIATION

given: latitude = 42°, S = 83%, α = 0.24, July 15,
T_{dp} = 9.3°C, T_a = 20.9°C, $T_{(max)}$ = 29.4°C, $T_{(min)}$ = 12.4°C.

find: R_a, R_s and R_n.

for July 15 the calendar day is D = 196 (non leap year)

Cooper declination model

$$\delta = 23.5\ \sin\left[\frac{360(D+284)}{365}\right] = 23.5\ \sin\left[\frac{360(196+284)}{365}\right] = 21.56° \qquad (2.34)$$

Showing four significant figures for δ is probably excessive.

Burman-Jacquot declination model

$$\delta = 0.3931 + 23.2577\ \cos\ (0.9861\ D - 170.7) + 0.3906\ \cos\ (1.9154\ D - 174.4)$$

$$\delta = 0.3931 + 23.2577\ \cos\ (0.9861 \cdot 196 - 170.7)$$

$$+0.3906\ \cos\ (1.9154 \cdot 196 - 174.4) = 21.50\ ° \qquad (2.35)$$

Using all three terms of the equation shown in Section 8.8.5 results in δ being equal to 21.54°.

Duffie-Beckman model sun-earth distance ratio

$$\left[\frac{\bar{d}}{d}\right]^2 = 1 + 0.034\cos\left[\frac{360\,D}{365.25}\right] = 1 + 0.034\cos\left[\frac{360 \cdot 196}{365.25}\right] = 0.9669 \qquad (2.36)$$

sunset hour angle

$$\cos\omega_{hs} = -\tan\delta\tan L = -\tan 21.56\tan 42 = -0.3558 \qquad (2.30)$$

$\omega_{hs} = 110.8\,°$

day length

$$H = \left[\frac{2\cdot 24}{360}\right]\omega_{hs} = \left[\frac{48}{360}\right]110.8 = 14.77\ hours \qquad (2.37)$$

extraterrestrial solar radiation R_a

$$R_a = \frac{SC}{\pi}\left[\frac{\bar{d}}{d}\right]^2 [\omega_{hs}\left\{\frac{\pi}{180}\right\}\sin L\sin\delta + \cos L\cos\delta\sin\omega_{hs}] \qquad (2.41)$$

$$R_a = \left[\frac{118.1}{\pi}\right][0.9669]\left[110.8\left\{\frac{\pi}{180}\right\}\sin 42\sin 21.56 + \cos 42\cos 21.56\sin 110.8\right] = 40.8\frac{MJ}{m^2\ day}$$

SC is the solar constant in MJ m^{-2} day^{-1} see Section 2.3.3.

solar radiation R_s

$$R_s = R_a\ (a + b\cdot S) = 40.8\ (0.25 + 0.5\cdot 0.83) = 27.1\frac{MJ}{m^2\ day} \qquad (2.57)$$

INSTANTANEOUS R_{so} MODELS

<u>ASHRAE model A</u>

given: 42°N, 1830m, July 15, solar noon

$$\cos\theta_z = \sin L\ \sin\delta + \cos L\ \cos\delta\cos\omega_h \qquad (2.28)$$

$$= \sin 42\ \sin 21.56 + \cos 42\ \cos 21.56\ \cos 0 = 0.9370$$

interpolate A, B and C using data from Table 2.7

Jun 21 = 172 (day number)
July 15 = 196
July 21 = 202

$A = 1088 + (1085 - 1088)\ (196 - 172)/(202 - 172) = 1086$

similarly
B = 0.207
C = 0.136
assume CN = 1

$\cos\theta_z = \sin\beta = 0.9370$

$\cos(\theta_z) = 1 / \sec(\theta_z)$

$$I_n = CN\ A\ \exp\left[-B \frac{P}{P_o} \sec\theta_z\right] \tag{2.44}$$

$$I_n = 1 \cdot 1086 \exp\left[-0.207 \frac{82.0}{101.3} \frac{1}{0.9370}\right] = 908.2 \frac{W}{m^2}$$

$$R_{so(d)} = \cos\theta_z\ I_n = 0.9370 \cdot 908.2 = 851 \frac{W}{m^2} \tag{2.44}$$

$$R_{so(s)} = C\ I_n = 0.136 \cdot 908.2 = 124 \frac{W}{m^2} \tag{2.46}$$

$$R_{so}(rate) = R_{so(d)} + R_{so(s)} = 851 + 124 = 975 \frac{W}{m^2} \tag{2.43}$$

<u>assume same conditions but 4PM local time and longitude is 106°</u>

$$local\ app.\ time = local\ std.\ time + longitude\ corr. + eq.\ of\ time \tag{2.31}$$

$$\Gamma = \frac{360}{365.242}(D - 1) = \frac{360}{365.242}(196 - 1) = 192.20 \tag{2.33}$$

$$EOT = 0.258 \cos\Gamma - 7.416 \sin\Gamma - 3.648 \cos 2\,\Gamma - 9.228 \sin 2\,\Gamma \tag{2.33}$$

$$EOT = -0.25 + 1.57 - 3.32 - 3.81 = -5.81\ minutes$$

$$longitude\ corr = 4(L_e - L_s) = 4(106 - 105) = 4\ minutes \tag{2.32}$$

$$local\ app.\ time = 4.00 + \tfrac{1}{60}(4.00 - 5.81) = 3.97 hours \quad (2.31)$$

$$360\ degrees = 24\ hours \qquad \therefore\ 15 degrees = 1\ hour$$

$$\omega = 3.97\ hours \cdot \frac{15^\circ}{hours} = 59.55^\circ$$

$$\cos\theta_z = \sin L\ \sin\delta + \cos L\ \cos\delta\cos\omega_h \quad (2.28)$$

$$= \sin 42\ \sin 21.56 + \cos 42\ \cos 21.56\ \cos 59.55 = 0.5962$$

$$I_n = CN\ A\ \exp\left[-B\frac{P}{P_o}\sec\theta_z\right] \quad (2.44)$$

$$I_n = 1 \cdot 1086\ \exp\left[-0.207\frac{82.0}{101.3}\frac{1}{0.5962}\right] = 819.9\ \frac{W}{m^2}$$

$$R_{so(d)} = \cos\theta_z\ I_n = 0.5962 \cdot 819.9 = 489\ \frac{W}{m^2} \quad (2.44)$$

$$R_{so(s)} = C\ I_n = 0.136 \cdot 819.9 = 112\frac{W}{m^2} \quad (2.46)$$

$$R_{so}(rate) = R_{so(d)} + R_{so(s)} = 489 + 112 = 601\ \frac{W}{m^2} \quad (2.43)$$

Reminder: The above value is for 4 P.M. local time.

ASHRAE model B

given: solar noon

$$\cos\theta_z = \sin\beta = 0.9370$$

$$I_{dn} = \frac{A}{\exp\left\{\frac{B}{\sin\beta}\right\}} = \frac{1086}{\exp\left\{\frac{0.207}{0.9370}\right\}} = 871\frac{W}{m^2} \quad (2.47)$$

$$R_{so(d)} = \cos\theta_z\ I_{dn} = 0.9370 \cdot 871 = 816\frac{W}{m^2} \quad (2.47)$$

$$R_{so(s)} = C\ I_{dn} = 0.136 \cdot 871 = 118\ \frac{W}{m^2} \tag{2.48}$$

$$R_{so}(rate) = R_{so(d)} + R_{so(s)} = 816 + 118 = 934\ \frac{W}{m^2} \tag{2.43}$$

daily R_{so} models

Heermann et al. (1985)

<u>exponential model</u>

$$\dot{A} = 31.25 + 0.001113\ z = 31.25 + 0.001113 \cdot 1830 = 33.29 \tag{2.50}$$

$$\dot{B} = 270 - 3.008\ L = 270 - 3.008 \cdot 42 = 143.7 \tag{2.51}$$

$$R_{so}(daily) = \dot{A}\ \exp-\left\{\frac{D-\dot{C}}{\dot{B}}\right\}^2 = 33.29\ \exp-\left\{\frac{196-170}{143.7}\right\}^2 = 32.2\frac{MJ}{m^2 day} \tag{2.49}$$

<u>cosine model</u>

$$A' = 31.54 - 0.2734\ L + 0.0007813\ z \tag{2.53}$$

$$A' = 31.54 - 0.2734 \cdot 42 + 0.0007813 \cdot 1830 = 21.49$$

$$B' = -0.2986 + 0.2678\ L + 0.0004102\ z \tag{2.54}$$

$$B' = -0.2986 + 0.2678 \cdot 42 + 0.0004102 \cdot 1830 = 11.70$$

$$R_{so}(daily) = A' + B'\cos\left\{\frac{2\pi D}{365} - C'\right\} \tag{2.52}$$

$$R_{so}(daily) = 21.49 + 11.70\cos\left[\frac{180}{\pi}\right]\left\{\frac{2\pi \cdot 196}{365} - 2.92\right\} = 32.0\frac{MJ}{m^2\ day}$$

note: the $180/\pi$ is a multiplier to convert radians to degrees

8.3.3 *Net radiation examples*

select: R_{so} = 32.2 MJ m^{-2} day^{-1}. Interpolated from Table 2.8 (and also estimated above)
R_s = 27.1 MJ m^{-2} day^{-1}
a, b, a_1 and b_1 from Table 2.11 for general conditions and arid climate

Net radiation is defined as the difference between the incoming and outgoing radiation including both the longwave and shortwave components. Therefore, in principle, R_n should be calculated using the most precise method. However, in practice, the principle of consistency is usually recommended and followed. Crop coefficients used to estimate ET_a depend on the method of estimating R_n recommended by the authors of the crop coefficients. For the most reliable estimates of ET, the recommended method of estimating R_n should be used.

Constant coefficients, Brunt equation

$$CR = R_s/R_{so} = 27.1/32.2 = 0.842 \qquad (2.65)$$

$$\varepsilon'_{bulk} = -[a(CR)+b][a_1 + b_1\sqrt{e_a}] \qquad (2.66)$$

$$= -[1.2\ (0.842) - 0.2][0.39 - 0.158\sqrt{0.903}] = -0.194$$

$T_{k(max)}$ = 302.6 K, $T_{k(min)}$ = 285.6 K

$$R_{nl} = \varepsilon'_{bulk}\ \sigma\left[\frac{T^4_{k(max)} + T^4_{k(min)}}{2}\right] \qquad (2.74)$$

$$= -0.194\ (4.9 \cdot 10^{-9})[\ (302.6^4 + 285.6^4)/2] = -7.15\frac{MJ}{m^2\ day}$$

$$R_n = (1-\alpha)R_s + R_{nl} = (1-0.23)27.1 - 7.15 = 13.7\frac{MJ}{m^2\ day} \qquad (2.61)$$

alternative estimate of ε'_{bulk}

$$\varepsilon'_{bulk} = -[a(CR)+b][-0.02 + 0.261\exp(-7.77 \cdot 10^{-4}(273 - T_k)^2)] \qquad (2.68)$$

$$\varepsilon'_{bulk} = -[1.2\ (0.842) - 0.2][-0.02 + 0.261\exp(-7.77 \cdot 10^{-4}(273 - 294.1)^2)] = -0.133$$

Variable coefficients, Brunt equation

$$a_1 = 0.26 + 0.1 \exp -[0.0154(30 \cdot m + N - 207)]^2 \qquad (2.70)$$

$$a_1 = 0.26 + 0.1 \exp -[0.0154(30 \cdot 7 + 15 - 207)]^2 = 0.353$$

$$b_1 = -0.139 \; always \qquad (2.71)$$

$$CR = R_s / R_{so} = 27.1/32.2 = 0.842 \qquad (2.65)$$

$$a = 1.126 \qquad b = -0.07 \qquad (2.72)$$

$$\varepsilon'_{bulk} = -[a(CR) + b][a_1 + b_1\sqrt{e_a}] \qquad (2.66)$$

$$\varepsilon'_{bulk} = -(1.126 \cdot 0.842 - 0.07)(0.353 - 0.139\sqrt{0.903}) = -0.194$$

$$R_n = R_s(1-\alpha) + R_{nl} = R_s(1-\alpha) + \varepsilon'_{bulk}\sigma\left[\frac{T^4_{k(\max)} + T^4_{k(\min)}}{2}\right] \qquad (2.74)$$

$$R_{nl} = -0.194 \cdot \frac{4.9}{10^9}\left[\frac{302.6^4 + 285.6^4}{2}\right] = -7.15 \frac{MJ}{m^2 \cdot day}$$

$$\alpha = 0.29 + 0.06 \sin[30(m + 0.0333\,N + 2.25)] \qquad (2.69)$$

$$\alpha = 0.29 + 0.06 \sin[30(7 + 0.0333 \cdot 15 + 2.25)] = 0.235$$

$$R_n = (1 - 0.235)\, 27.1 - 7.15 = 13.6 \frac{MJ}{m^2 \cdot day} \qquad (2.67)$$

8.4 Chapter 4 examples

The following examples are presented to illustrate methods of estimating *reference crop evapotranspiration*, ET_r. Methods discussed here use procedures of chapter 4.

8.4.1 *Input data & parameters*

given:
$T_{max} = 29.4°C$ Elevation = 1950m

T_{min} = 8.9°C Latitude = 42°
T_{dp} = 6.6°C July 15
R_s = 24.8 MJ m^{-2} day^{-1} $\alpha = 0.24$
u_2 = 201 km day_{-1}
assume G = 0.0 MJ m^{-2} day^{-1} see Section 2.6 for more discussion of this assumption.
Using methods shown in chapter 2. Parameters may differ slightly because of round-off and other details.
λ = 2456 kJ kg^{-1} = 2.456 MJ kg^{-1}
γ = 0.0534 kPa K^{-1}
Δ = 0.1382 kPa K^{-1}

8.4.2 *Constant parameters*

Jensen-Haise seasonal parameters

The instructions found in various publications involving the Jensen-Haise method, for example Jensen et al. (1990), must be followed with care. Saturation vapor pressures are at the mean maximum and mean minimum monthly air temperatures for the warmest month of the year, according to long term climatic data.

Given:
Avg. monthly maximum air temperature = 28.4°C
Avg. monthly minimum air temperature = 7.7°C.

$$e_2^o = 0.6108 \exp\left[\frac{17.27 \cdot 28.4}{28.4 + 237.3}\right] = 3.87\ kPa \qquad (2.14)$$

$$e_1^o = 0.6108 \exp\left[\frac{17.27 \cdot 7.7}{7.7 + 237.3}\right] = 1.05\ kPa \qquad (2.14)$$

$$C_H = \frac{5.0}{e_2 - e_1} = \frac{5.0}{3.87 - 1.05} = 1.773 \qquad (4.56)$$

$$C_1 = 38 - \frac{z}{152.5} = 38 - \frac{1950}{152.5} = 25.2 \qquad (4.57)$$

$$C_T = \frac{1}{C_1 + 7.3\, C_H} = \frac{1}{25.2 + 7.3 \cdot 1.773} = \frac{1}{25.2 + 7.3 \cdot 1.773} = 0.0262 \qquad (4.55)$$

$$T_x = -2.5 - 1.4\,(e_2 - e_1) - z/550 \qquad (4.58)$$

$$= -2.5 - 1.4\,(3.87 - 1.05) - 1950/550 = -9.99\ ^\circ C$$

See Section 8.4.8 for an example of a monthly calculation using the Jensen-Haise equation.

8.4.3 *Original Penman method*

Alternate methods of estimating f(e) have been used. Two methods of estimating f(e) or VPD appear.

$$\lambda \mathrm{ET}_{rg} = \frac{\Delta}{\Delta+\gamma}(R_n + G) + \frac{\gamma}{\Delta+\gamma} 6.43\, f(e)\, f(u) \tag{4.14}$$

$$e^o = 0.6108 \exp\left[\frac{17.27\, T}{T+237.3}\right] \tag{2.14}$$

$$e^o_{\max} = 0.6108 \exp\left[\frac{17.27 \cdot 29.4}{29.4+237.3}\right] = 4.10\ kPa$$

$$e^o_{avg} = 0.6108 \exp\left[\frac{17.27 \cdot 19.2}{19.2+237.3}\right] = 2.22\ kPa$$

$$e^o_{\min} = 0.6108 \exp\left[\frac{17.27 \cdot 8.9}{8.9+237.3}\right] = 1.14\ kPa$$

$$e_a = 0.6108 \exp\left[\frac{17.27 \cdot 6.6}{6.6+237.3}\right] = 0.97\ kPa$$

$$VPD_1 = e^o_{T_a} - e^o_{T_{dp}} = e^o_{T_{avg}} - e^o_a = 2.22 - 0.97 = 1.25\ kPa \tag{4.15}$$

$$R_{nl} = -[a\ (R_s/R_{so}) + b]\,[a_1 + b_1\sqrt{e_a}]\,\sigma[(T^4_{k(\max)} + T^4_{k(\min)})/2)] \tag{4.19}$$

select: a = 1.22, b = -0.18, a_1 = 0.325 and b_1 = -0.139 from Table 2.11 assuming southern Idaho conditions apply

$$T_{k(\max)} = 29.4 + 273.15 = 302.6\ K \qquad T_{k(\min)} = 8.9 + 273.15 = 282.0\ K$$

the following value of R_{so} was selected by inspection from Table 2.8

$$R_{so} = 32.2\ \frac{MJ}{m^2\ day} \qquad \frac{R_s}{R_{so}} = \frac{24.8}{32.2} = 0.770$$

$$R_{nl} = -[1.22 \cdot 0.770 - 0.18]\,[0.325 - 0.139\sqrt{0.97}\,]\left\{\frac{4.9}{10^9}\right\}[302.6^4 + 282.0^4]/2 = -5.1\ \frac{MJ}{m^2 day}$$

The albedo, α, can be constant or it can vary throughout the growing season. The value of 0.24 is selected because it is given as part of this example. See Sections 2.3.4 through 2.3.6 for a more complete discussion of α.

$$R_n = (1-\alpha)R_s + R_{nl} = (1-0.24)\ 24.8 - 5.1 = 13.7\ \frac{MJ}{m^2 day}$$

$$f(u) = 1 + 0.0062\ u_2 = 1 + 0.0062 \cdot 201 = 2.25 \qquad (4.16)$$

$$\lambda \mathrm{ET}_{rg} = \frac{0.1382}{0.1382+0.0534}\ 13.7 + \frac{0.0534}{0.1382+0.0534}\ 6.43 \cdot 1.25 \cdot 2.25 = 14.9\ \frac{MJ}{m^2 day} \qquad (4.14)$$

$$\mathrm{ET}_{rg} = \frac{14.9}{2.456} = 6.1\ \frac{mm}{day}$$

Jensen et al. (1990) show an example of ET being calculated using VPD_2.

$$f(e) = VPD_2 = \frac{e^o_{\max} + e^o_{\min}}{2} - e_a = \frac{4.10+1.14}{2} - 0.97 = 1.65\ kPa \qquad (4.11)$$

$$\lambda \mathrm{ET}_{rg} = \frac{0.1382}{0.1382+0.0534}\ 13.7 + \frac{0.0534}{0.1382+0.0534}\ 6.43 \cdot 1.65 \cdot 2.25 = 16.5\ \frac{MJ}{m^2 day}$$

$$\mathrm{ET}_{rg} = \frac{16.5}{2.456} = 6.7\ \frac{mm}{day}$$

8.4.4 *Penman equation, Wright and Jensen 1972 examples*

$$\lambda \mathrm{ET}_{ra} = \frac{\Delta}{\Delta+\gamma}(R_n + G) + \frac{\gamma}{\Delta+\gamma} 6.43 f(e) f(u) \qquad (4.17)$$

$$f(u) = 0.75 + 0.0115 u_2 \qquad (4.20)$$

$$f(u) = 0.75 + 0.0115 \cdot 201 = 3.06$$

$$f(e) = VPD_2 = \frac{\left[e^o_{\max} + e^o_{\min}\right]}{2} - e_a \qquad (4.11)$$

$$f(e) = \frac{4.10+1.14}{2} - 0.97 = 1.65\ kPa$$

See the previous section for calculation details for R_n = 13.7 MJ m^{-2} day^{-1}.

$$\lambda \mathrm{ET}_{ra} = \frac{0.1382}{0.1382+0.0534}13.7 + \frac{0.0534}{0.1382+0.0534}\cdot 6.43\cdot 1.65\cdot 3.06 = 18.9\,\frac{MJ}{m^2\,day}$$

$$\mathrm{ET}_{ra} = \frac{18.9}{2.456} = 7.7\,\frac{mm}{day}$$

8.4.5 *Penman equation, Wright 1982 examples*

$$\lambda \mathrm{ET}_{ra} = \frac{\Delta}{\Delta+\gamma}(R_n + G) + \frac{\gamma}{\Delta+\gamma}6.43 f(e) f(u) \quad (4.21)$$

the following value of R_{so} was selected by inspection from Table 2.8

$$R_{so} = 32.2\,\frac{MJ}{m^2\,day} \qquad \frac{R_s}{R_{so}} = \frac{24.8}{32.2} = 0.770$$

$$if \quad \frac{R_s}{R_{so}} > 0.7 \quad then$$

$$\alpha = 0.29 + 0.06\sin[30(m + 0.0333\,N + 2.25)] \quad (4.23)$$

$$\alpha = 0.29 + 0.06\sin[30(7 + 0.0333\cdot 15 + 2.25)] = 0.235$$

$$a_1 = 0.26 + 0.1\,\exp\{-[0.0154(30m + N - 207)]^2\} \quad (4.27)$$

$$a_1 = 0.26 + 0.1\,\exp\{-[0.0154(30\cdot 7 + 15 - 207)]^2\} = 0.353$$

$$\frac{R_s}{R_{so}} > 0.7 \quad a = 1.126 \quad b = -0.07 \quad (4.25)$$

$$T_{k(\max)} = 29.4 + 273.15 = 302.6\,K \qquad T_{k(\min)} = 8.9 + 273.15 = 282.0\,K$$

$$R_{nl} = -[a\cdot R_s/R_{so} + b]\,[a_1 + b_1\sqrt{e_a}]\,\sigma[(T^4_{k(\max)} + T^4_{k(\min)})/2)] \quad (4.24)$$

$$R_{nl} = -[1.126\cdot 0.770 - 0.07]\,[0.353 - 0.139\,\sqrt{0.97}\,]\left\{\frac{4.9}{10^9}\right\}\,[302.6^4 + 282.0^4]/2 = -6.2\,\frac{MJ}{m^2 day}$$

$$R_n = (1-\alpha)R_s + R_{nl} = (1 - 0.235)\,24.8 - 6.2 = 12.8\,\frac{MJ}{m^2 day} \quad (4.22)$$

$$a_w = 0.4 + 1.4\exp\{-[(D-173)/58]^2\} \qquad (4.30)$$

$$a_w = 0.4 + 1.4\exp\{-[(196-173)/58]^2\} = 1.60$$

$$b_w = 0.007 + 0.004\exp\{-[(D-243)/80]^2\} \qquad (4.31)$$

$$b_w = 0.007 + 0.004\exp\{-[(196-243)/80]^2\} = 0.0098$$

$$f(u) = a_w + b_w\, u_2 = 1.60 + 0.0098 \cdot 201 = 3.57 \qquad (4.29)$$

$$\lambda \mathrm{ET}_r = \frac{0.1382}{0.1382+0.0534}(12.8+0.0) + \frac{0.0534}{0.1382+0.0534} 6.43 \cdot 1.65 \cdot 3.57 = 19.8 \frac{MJ}{m^2\, day}$$

$$\mathrm{ET} = \frac{19.8}{2.456} = 8.1 \frac{mm}{day}$$

Note: R_n was especially calculated for this example because α varies with date in the Wright (1982) Penman method.

8.4.6 *FAO Penman*

Additional information
Day-night wind ratio = 1.70
RH_{max} = 87%
RH_{min} = 22%

$$\mathrm{ET}_{rg} = c\left[\frac{\Delta}{\Delta+\gamma}\left\{\frac{H}{\lambda}\right\} + \frac{\gamma}{\Delta+\gamma} 2.7 f(u) f(e)\right] \qquad (4.32)$$

$$\overline{RH} = \frac{(87+22)}{2} = 54.5\%$$

$$\overline{T} = \frac{(29.4+8.9)}{2} = 19.2°C$$

$$e^o_{T_a} = 0.6108\,\exp\left[\frac{17.27 \cdot 19.2}{19.2+237.3}\right] = 2.22\ kPa \qquad (2.14)$$

$$f(e) = (1-\overline{RH})\, e^o_{T_a} = (1-0.545)\ 2.22 = 1.01\ kPa \qquad (4.35)$$

$$f(u) = (1+0.01\, u_2) = (1+0.01 \cdot 201) = 3.01 \qquad (4.34)$$

$$U_d = \left[\frac{DN_r \cdot u_2}{1+DN_r}\right]\left[\frac{1000}{12 \cdot 3600}\right] = \left[\frac{1.70 \cdot 201}{1+1.70}\right]\left[\frac{1000}{12 \cdot 3600}\right] = 2.93\ \frac{m}{\text{sec}} \tag{4.6}$$

$$\frac{R_s}{\lambda} = \left[\frac{24.8}{2.456}\right] = 10.1\ \frac{mm}{day}$$

see Section 8.2.3 for an explanation of this conversion. In FAO-24 (Dorrenbos and Pruitt, 1977) R_n has the units of mm day^{-1} necessitating the conversion appearing above. This leads to an inconsistency in units.

$$c = a_o + a_1 RH_{\max} + a_2 R_s + a_3 U_d + a_4 DN_r + a_5 U_d DN_r + a_6 RH_{\max} R_s U_d$$

$$+ a_7 RH_{\max} R_s DN_r \tag{4.37}$$

$$c = 0.6817006 + 0.0027864 \cdot 87 + 0.0181768 \cdot 10.1 - 0.0682501 \cdot 2.93 + 0.0126514 \cdot 1.7 + 0.0097297 \cdot 2.93 \cdot 1.7$$

$$+ 0.000043025 \cdot 87 \cdot 10.1 \cdot 2.93 - 0.000000092118 \cdot 87 \cdot 10.1 \cdot 1.7$$

$$c = 0.6817 + 0.2424 + 0.1836 - 0.2000 + 0.0215 + 0.0485 + 0.1108 - 0.0001 = 1.088$$

FAO-24 has different constants for the determination of R_n but in general follows the procedures shown in Section 2.5.3. The determination of R_n should be by the best possible method but there is much to be said for following the recommendations in FAO-24 in terms of consistency. However for convenience and the fact that the constants used in Section 8.4.3 are more appropriate for local conditions use R_n as calculated in Section 8.4.3.

$$ET_{rg} = 1.088\left[\frac{0.1382}{0.1382+0.0534}\left\{\frac{13.7}{2.456}\right\} + \frac{0.0534}{0.1382+0.0534} \cdot 2.7 \cdot 3.01 \cdot 1.01\right] = 6.9\ \frac{mm}{day}$$

8.4.7 *Penman-Monteith equation reference ET examples*

given:
same data as for the original Penman plus
crop height, *alfalfa*, 0.39 m
wind measurement height 2 m
temperature and humidity measurement heights 1.5 m

$$\lambda\, ET = \frac{\Delta \cdot H + \rho\, C_p\, Q}{\Delta + \gamma^*} \tag{4.38}$$

$$Q = \frac{e^o_{(T_z)} - e_z}{r_a} = \frac{kPa \cdot m}{\sec} \tag{4.39}$$

Penman-Monteith wind term units

$$\frac{\rho C_p Q}{\Delta + \gamma^*} = \left[\frac{kg}{m^3}\right]\left[\frac{kJ}{kg \cdot K}\right]\left[\frac{kPa \cdot m}{\sec}\right]\left[\frac{K}{kPa}\right]\left[\frac{86400 \sec}{day}\right]\left[\frac{1000}{k}\right]\left[\frac{M}{10^6}\right] = \frac{MJ}{m^2\, day}$$

calculations

$$LAI = 5.5 + 1.5 \ln(z_v) = 5.5 + 1.5 \ln(0.39) = 4.09 \tag{4.52}$$

$$z_{om} = 0.123\, z_v = 0.123 \cdot (0.39) = 0.048\, m \tag{4.48}$$

$$z_{ov} = 0.1\, z_{om} = 0.1 \cdot 0.048 = 0.0048\, m \tag{4.49}$$

$$d = 0.67\, (z_v) = 0.67\, (0.39) = 0.261\, m \tag{4.50}$$

$$u_{2(m/s)} = \left[\frac{201\, km}{day}\right]\left[\frac{day}{86400 \sec}\right]\left[\frac{1000}{k}\right] = 2.33\, \frac{m}{\sec}$$

$$r_a = \ln\{(z_w - d)/z_{om}\} \ln\{(z_h - d)/z_{oh}\}/k^2 u_{2(m/s)} \tag{4.47}$$

$$r_a = \frac{\ln\left[\frac{2.0 - 0.261}{0.0480}\right] \ln\left[\frac{1.5 - 0.261}{0.00480}\right]}{0.41^2 \cdot 2.33} = 50.9\, \frac{\sec}{m}$$

$$r_c = \frac{r_l}{0.5\, LAI} = \frac{100}{0.5 \cdot 4.09} = 48.9\, \frac{\sec}{m} \tag{4.51}$$

$$\gamma^* = \gamma\left[1 + \frac{r_c}{r_a}\right] = 0.0534\left[1 + \frac{48.9}{50.9}\right] = 0.1047\, \frac{kPa}{K} \tag{4.41}$$

$$Q = \frac{e^o_{(T_z)} - e_z}{r_a} = \frac{1}{r_a}\left[\frac{e^o_{\max} + e^o_{\min}}{2} - e_a\right] \tag{4.39}$$

$$Q = \frac{1}{50.9}\left[\frac{4.10 + 1.14}{2} - 0.97\right] = 0.0324\, \frac{kPa \cdot m}{\sec}$$

use ρ as 0.976 kg m^{-3} from the example calculations found in Section 8.3.2. It can be argued that ρ should be calculated for this specific example. See Sections 2.2.1 and 8.3.2 for details about the calculation of ρ. An average value of C_p is used, see Section 8.7.6 for justification of this assumption. For convenience use R_n = 13.7 MJ m^{-2} day^{-1} as calculated in Section 8.4.3.

$$\lambda ET = \frac{0.1382 \cdot 13.7}{0.1382 + 0.1047} + \frac{0.976(1.01)(1000)(0.0324)\,86400}{(0.1382 + 0.1047)\,10^6} = 19.2 \frac{MJ}{day\,m^2} \qquad (4.38)$$

$$ET = \frac{19.2}{2.456} = 7.8 \frac{mm}{day}$$

8.4.8 *Jensen-Haise method*

$$\lambda\, ET_{ra} = C_T(T - T_x)\, R_s \qquad (4.54)$$

$$ET_{ra} = C_T(T - T_x)\frac{R_s}{\lambda} = 0.0262\,(19.2 - -9.99)\;10.1 = 7.7 \frac{mm}{day}$$

See Section 8.4.2 for an example showing the calculation of C_T and T_x. These values do not change within the growing season.

8.4.9 *FAO-24 radiation method (Makkink)*

given:
same data as for FAO-24 Penman method

$$ET_{rg} = a + b\left[\frac{\Delta}{\Delta + \gamma}\right]\frac{R_s}{\lambda} \qquad (4.60)$$

$$\overline{RH} = \frac{87 + 22}{2} = 54.5\ \%$$

$$\frac{R_s}{\lambda} = \frac{24.8}{2.456} = 10.1 \frac{mm}{day}$$

$$U_d = 2.93 \frac{m}{sec}$$ *from FAO Penman example*

$$a = -0.3 \qquad (4.61)$$

$$b = a_o + a_1 \cdot \overline{RH} + a_2 \cdot U_d + a_3 \cdot \overline{RH} \cdot U_d + a_4 \cdot \overline{RH}^2 + a_5 \cdot U_d^2 \qquad (4.62)$$

$$b = 1.0656 - 0.0012795 \cdot 54.5 + 0.044953 \cdot 2.93 - 0.0002033 \cdot 54.5 \cdot 2.93 - 0.000031508 \cdot 54.5^2 - 0.0011026 \cdot 2.93^2$$

$$b = 1.066 - 0.070 + 0.132 - 0.032 - 0.094 - 0.009 = 0.993$$

$$\mathrm{ET}_{rg} = a + b\left[\frac{\Delta}{\Delta + \gamma}\right]\frac{R_s}{\lambda} = -0.3 + 0.993\left[\frac{0.1382}{0.1382 + 0.0534}\right] 10.1 = 6.9 \frac{mm}{day} \qquad (4.60)$$

8.4.10 *FAO-24 Blaney-Criddle method examples*

Same information as for previous examples
select p from Table 2.4 using interpolation

fundamental equations

$$\mathrm{ET}_{rg} = a + bf \qquad (4.66)$$

$$f = p(0.46\,T + 8.13) \qquad (4.67)$$

$$a = 0.0043 RH_{\min} - n/N - 1.41 \qquad (4.68)$$

$$b = a_o + a_1 RH_{\min} + a_2 n/N + a_3 U_d + a_4 RH_{\min} n/N + a_5 RH_{\min} U_d \qquad (4.69)$$

calculations

interpolating using latitude and Table 2.4

$$p_{42} = 10.38 + (11.24 - 10.38)\frac{(42-40)}{(50-40)} = 10.55\ \%$$

according to the method p must be on a daily basis

$$p = \frac{10.55}{31} = 0.3403\ \%$$

According to Equation 2.55 n/N = S and using R_s = 24.8 MJ m^{-2} day^{-1}
from Table 2.5 using interpolation, R_a = 40.8 MJ m^{-2} day^{-1}.

$$R_s = R_a\,(0.25 + 0.5\,n/N) \qquad (2.57)$$

$$24.8 = 40.8\,(0.25 + 0.5\,n/N) \qquad n/N = 0.72$$

$$f = p\ (0.46\ T + 8.13) = 0.3403\ (0.46 \cdot 19.2 + 8.13) = 5.77 \qquad (4.67)$$

$$RH_{\min} = 100 \cdot \frac{0.97}{4.10} = 23.7\ \%$$ *alternative determination*

$$a = 0.0043\ RH_{\min} - n/N - 1.41 = 0.0043 \cdot 23.7 - 0.72 - 1.41 = -2.02 \qquad (4.68)$$

$$b = a_o + a_1 RH_{\min} + a_2 n/N + a_3 U_d + a_4 RH_{\min} n/N + a_5 RH_{\min} U_d \qquad (4.69)$$

$$b = 0.81917 - 0.0040922 \cdot 23.7 + 1.0705 \cdot 0.72 + 0.065649 \cdot 2.93 - 0.0059684 \cdot 23.7 \cdot 0.72$$

$$-0.0005967 \cdot 23.7 \cdot 2.93$$

$$b = 0.8192 - 0.0969 + 0.7708 + 0.1924 - 0.1018 - 0.0414 = 1.54$$

$$\mathrm{ET}_{rg} = a + b\ f = -2.02 + 1.54 \cdot 5.77 = 6.9\ \frac{mm}{day} \qquad (4.66)$$

8.5 Chapter 5 examples

given:
spring wheat at Seedskadee Wyoming (Pochop and Burman, 1987) with a growing season of May 1 through August 15 with a distribution of 13 days for the initial stage, 22 days for the developmental stage, 45 days for the mid-season stage and 27 days for the late-season stage. This is a total of 107 days. Full cover was observed to occur on May 28. The following are based upon local information and experience.

This information indicates an observed short growing season associated with high altitude conditions in the Rocky Mountain area of the U.S.

Table 8.1
Input data for example calculations
from Pochop and Burman (1987)

	Meas ET	est ET_r Kincaid Heermann (1974)	% available soil water
May	136	212	85
June	162	226	82
July	195	232	85
Aug	168	191	94

ET amounts are in mm per month

8.5.1 *FAO-24 Crop Coefficients*

The first task is to construct the crop curve using the data and instructions found in FAO-24 (Doorenbos and Pruitt, 1977). Much of the construction involves arbitrary selections. Exact interpretations of procedures involves judgment due to the nature of instructions contained in FAO-24. The following assumptions were made.

1. Growing season dates were selected based upon the observation of the crop and consultation with individuals familiar with the area and crop production.

2. A K_c value for the initial stage of 0.4 was selected from Figure 5.6 based upon an interval of rain or irrigation of about 9 days and an ET_r of approximately 4.5 mm per day.

3. A K_c value of 1.2 was selected for the mid-season stage based upon the interpretation that the area represents the higher wind conditions and the lower humidity conditions found in Table 5.8 rather than the use of specific numerical values of minimum humidity and wind velocity. FAO-24 does not specifically say that average or maximum conditions should be used.

4. A K_c of 0.2 was selected for the late season stage from Table 5.8 for

the same reasoning as described above.

Figure 8.1 shows the crop coefficient curve for spring wheat at Seeds-kadee Wyoming, USA.

8.5.2 *ASCE Crop Coefficients*

The following example may be considered to be unorthodox because the mean alfalfa based crop coefficients found in Tables 5.11 through 5.16 are combined with the growing season information found in FAO-24 (Jensen et al., 1990: Doorenbos and Pruitt, 1977). See the previous example for input data and comparisons of methods.

Calculation procedures

The same growing season dates and crop as the previous example are assumed. For the beginning of the *development* stage the date is May 14 and the calendar day is number 134. Other day numbers are explained in Table 8.2.

$$100\left[\frac{134-121}{148-121}\right]=48\ \%$$

from Table 5.11 K_c is interpolated to be 0.52.

At the beginning of the *late season* growth stage the day number is 201 corresponding to July 20.

$$(201-148)=53\ days$$

using Table 5.12 K_c is interpolated to be 0.44.

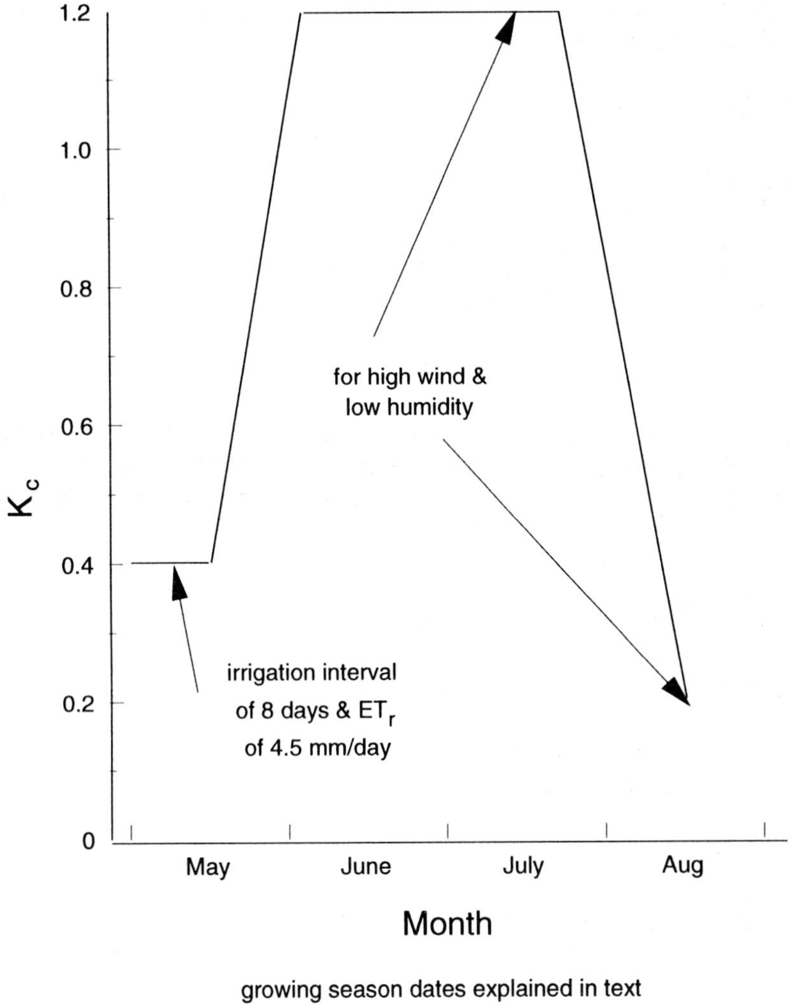

Figure 8.1
Crop curve constructed for spring wheat
FAO-24 methods, Seedskadee Wyoming
straight lines may be rounded if desired

Table 8.2
Illustration of average K_c values
Jensen et al. (1990); Wright (1982)

interval or harvest	days interval	start date	calendar day	% time or days	K_c mean
initial	13	May-1	121	0	0.20
intermediate	22	May-14	134	48	0.52
mid-season	45	June-5	156	8	1.00
late-season	27	July-20	201	53	0.44
harvest	0	Aug-15	227	79	0.11

full cover is May 28
day number 148

Graphical crop curve

The crop curve is constructed by using the same vertical lines which define the start and stop of each growing season interval. The points are plotted on each line of the preceding table in the same manner as constructing an FAO crop curve.

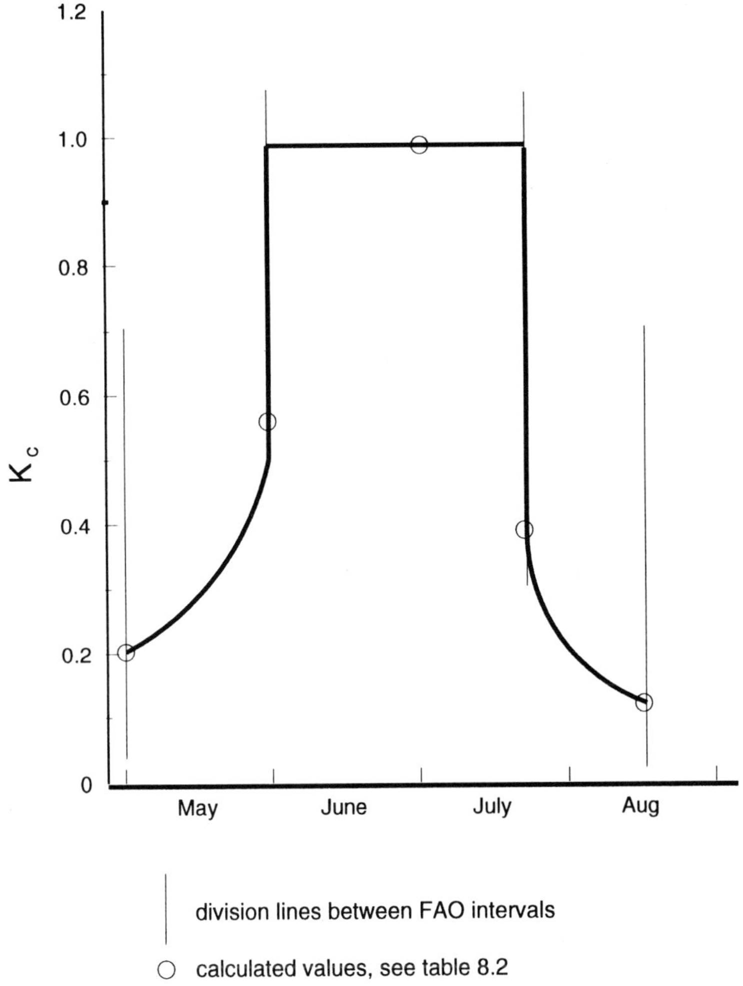

Figure 8.2
Crop curve constructed from mean values
Jensen et al. (1990); Wright (1982)
straight lines may be curved if desired

given:
Determine the average crop coefficient for each of the months from May through August 15. Then apply the natural logarithmic correction shown in

equation 5.7 to determine the ET for the season. Use the ET calculated by the Penman method calibration of Kincaid and Heermann (1974). These calculations are summarized in the following table:

Table 8.3
Growing season ET including soil water corrections

month	K_c	monthly $KH\text{-}ET_r$	% AW	SW correction	monthly ET
May	0.32	212	85	0.965	65
June	1.00	226	82	0.958	217
July	0.72	232	85	0.965	161
Aug	0.17	191	94	0.987	16

ET amounts are in mm per month

$$K_a = \frac{\ln(\%PESW + 1)}{\ln(101)} = \frac{\ln(85 + 1)}{\ln(101)} = 0.965 \tag{5.7}$$

$$ET_a = 0.32 \cdot 212 \cdot 0.965 = 65 \frac{mm}{month}$$

8.6 Chapter 7 examples

Making evaporation estimates from free water surfaces is given different treatment because of significant differences. In many cases it is very difficult from a practical standpoint to convert the equations so strict SI units may be used. Any test data is of necessity different. The problem is different from a conceptual standpoint (see Chapters 4 and 7) even though evaporation and evapotranspiration are based on Dalton's law.

8.6.1 *Evaporation units and input data*

The SI System of units is followed as closely as possible except in the discussion of example calculations involved in making evaporation estimates. Some of the following material has units such as *cm* even though the centimeter is not a recommended SI unit. The following methods were calibrated using archaic units, and some of the more complex equations have units which would be difficult to convert to true SI units from a practical standpoint. Therefore, units such as *cm* have been retained primarily from the viewpoint of accuracy. The Kohler-Nordenson-Fox equation would be very difficult to change to strict SI units.

An additional reason for retaining non-SI units is the fact that it is easy to make unit conversions of input data to any desired units in any computer program used to make evaporation estimates.

Monthly averages of climatic data from Laramie, Wyoming for August 1987 were selected.

Given:
elevation 2200 m
latitude 41° 19' north
longitude 105° 36' W
$u_p = 261$ km day^{-1}
$R_s = 2147$ J cm^{-2} day^{-1}
RH = 43.7%
$T_a = 15.2$ °C
measured $E_p = 7.76$ mm day^{-1} (for class A pan)

8.6.2 *General parameters*

The following general parameters are similar to those found in the preceding material. Units and, hence, calculation methods differ.

Atmospheric pressure

$$P = 101.3 - 0.01055 \cdot 2200 = 78.1\ kPa \qquad (2.9)$$

Saturation vapor pressure

$$e^o = 0.6108 \cdot \exp\left[\frac{17.27 \cdot 15.2}{237.3 + 15.2}\right] = 1.727\ kPa \qquad (2.14)$$

Actual vapor pressure

$$e_a = 0.437 \cdot 1.727 = 0.755\ kPa \quad (2.16)$$

Dew point temperature

using an equation solver or trial and error and Equation 2.14
T_{dp} = 3.0 °C

8.6.3 *Kohler-Nordenson-Fox equation*

Psychrometric constants

$$\gamma_p = 0.001568 \cdot P = 0.001568 \cdot 78.1 = 0.1225 \frac{kPa}{^\circ C} \quad (7.2)$$

$$\gamma_l = 0.000661 \cdot P = 0.000661 \cdot 78.1 = 0.0516 \frac{kPa}{^\circ C} \quad (7.6)$$

Slope of saturated vapor pressure versus temperature curve

$$\Delta = \frac{(e_s - e_a)}{(T_a - T_{dp})} = \frac{(1.727 - 0.755)}{(15.2 - 3.0)} = 0.0797 \frac{kPa}{^\circ C} \quad (7.5)$$

Aerodynamic function

$$E_a = 25.4[0.296((e_s - e_a)^{0.88}\}(0.37 + 0.00255u_p)] \quad (7.3)$$

$$E_a = 25.4\,[0.296\{(1.727 - 0.755)^{0.88}\}\,(0.37 + 0.00255 \cdot 261)] = 7.60 \frac{mm}{day}$$

Net radiation

$$R_n\Delta = 154.4\exp[(1.8T_a - 180)(0.1024 - 0.01066 \cdot \ln(0.239R_s)) - 0.01544] \quad (7.4)$$

$$R_n\Delta = 154.4\exp[(1.8 \cdot 15.2 - 180)(0.1024 - 0.01066 \cdot \ln(0.239 \cdot 2147)$$

$$-0.01544] = 0.6364$$

Pan evaporation

$$E_p = \frac{\Delta R_n + \gamma_p E_a}{\Delta + \gamma_p} \tag{7.1}$$

$$E_p = \left[\frac{0.6364 + 0.1225 \cdot 7.6}{0.0797 + 0.1225} \right] = 7.8 \frac{mm}{day}$$

Lake evaporation

$$E_l = 0.7 \left[\frac{(\Delta R_n + \gamma_l E_a)}{(\Delta + \gamma_l)} \right] \tag{7.7}$$

$$E_l = 0.7 \left[\frac{0.6364 + 0.0516 \cdot 7.6}{0.0797 + 0.0516} \right] = 5.5 \frac{mm}{day}$$

8.7 Property standards

Traditionally, the Smithsonian Meteorological Tables (List, 1984) have been accepted as providing standard values of thermodynamic properties for the lower atmosphere. Thermodynamic property tables known as data *steam tables* also provide data on properties of liquid water, water vapor, and mixtures of both liquid water and water vapor. One of the most common references in thermodynamic literature is Keenan, et al., (1979). Data shown in steam tables may also be assumed to represent standard values because of their wide acceptance and revisions over a period of many years.

Is it correct to use steam table data for psychrometric purposes? Silver and Nydahl (1977) use steam table data in psychrometric example problems. Van Wylen and Sonntag (1985) discuss the use of enthalpy information in psychrometric charts. Data common to both the Smithsonian Meteorological Tables and the steam tables agree very closely. Both sources will be used in the following and referred to as being standards.

8.7.1 *Atmospheric pressure*

Average atmospheric pressure is needed to calculate the psychrometer constant and air density. For routine estimates of ET, an average value for a location is sufficient. Changes in atmospheric pressure because of the passage of storm fronts are not important in estimating ET.

Equations 7.2a and 7.2b in *Consumptive Use of Water and Irrigation Water Requirements* (Jensen, 1974) were developed by Burman by plotting values of atmospheric pressure against elevation using the Smithsonian Meteorological Tables (List, 1984). These tables represent the National

Advisory Committee for Aeronautics (NACA) Standard Atmosphere. The resulting plot yielded a straight line relationship between atmospheric pressure and elevation with minimal deviations. The slope of the line was determined graphically, and the intercept was set at 1013 mb for sea level conditions. These data represent standard conditions at mid-latitudes in the US. Equations 7.2 and 7.2b are:

$$P = 1013 - 01055z, \quad z \text{ in } m \tag{8.1}$$

$$P = 1013 - 0.03217\,z, \quad z \text{ in } ft. \tag{8.2}$$

where z is *elevation* above sea level, and *atmospheric pressure* P is in mb.

These equations are easy to use but are difficult to explain without referring to the Standard Atmosphere and basic thermodynamics. The user has no method of adjusting estimates for use in a different location. The procedure shown below can easily be modified by using different base conditions and lapse rates.

The hydrostatic equation is shown below:

$$\frac{dP}{dZ} = -\rho g \tag{8.3}$$

where P is *atmospheric pressure* in kPa, z is *elevation* in meters, g is *acceleration gravity* in m s^{-2}, and ρ is *density* in kg m^{-3}.

Commonly four integrations of the hydrostatic equation for the atmosphere are described (Hess, 1959). These are the homogeneous atmosphere, the isothermal atmosphere, the constant lapse rate atmosphere, and the dry adiabatic atmosphere. The homogeneous assumption is not realistic and only serves as a comparison.

The portion of the atmosphere known as the stratosphere is from 10.8 to 32km is a zone of constant temperature assumed to be at a temperature of -55 degrees C and is hence an *isothermal* atmosphere. The stratosphere is of no interest in estimating ET.

The constant lapse rate assumption fits actual conditions in the lower atmosphere from sea level to an elevation of approximately 11 km. Fox and McDonald (1985) present an excellent diagram of the US Standard Atmosphere. The constant lapse rate model results from an integration of a combination of the hydrostatic equation, the constant lapse rate, and the ideal gas law. The ideal gas law represents low density and low temperature gases. Atmospheric air fits the ideal gas law very well. The result of the integration follows:

$$\frac{P}{P_o} = \left[\frac{T_o - \alpha(z - z_o)}{T_o} \right]^{\frac{g}{\alpha R}} \tag{8.4}$$

where P is *atmospheric pressure* in kPa, α is the *lapse rate* in °C or K km^{-1}, z is *elevation* in km, T represents air temperature in K, *g* is *acceleration of gravity* in m s^{-2}, and R is the *specific gas constant* for dry air kJ kg^{-1} K^{-1}. The subscript *o* refers to a reference level, usually sea level. Equation 8.4 represents the lower portion of the US Standard Atmosphere. The use of the specific gas constant for dry air is not theoretically defensible but is justified practically because the average atmospheric pressure may be estimated reliably using Equation 8.4.

Equation 8.4 can be used to estimate an average annual atmospheric pressure for the elevation selected, based upon the assumed reference level, seasonal temperature, and the lapse rate selected. The US Standard Atmosphere assumes T_0 is 288K, P_0 is 101.3 kPa, and α is -6.5 °C km^{-1} for mid-latitudes. If an average pressure and temperature are known for a base elevation other than sea level, these may be used for T_0, P_0 and z_0. In addition, a local lapse rate α may be used.

Table 8.1 is a comparison of values of estimated atmospheric pressure with typical random measurements. Differences in pressure of these magnitudes will not lead to an error of practical significance in estimating ET. Typically, variations in atmospheric pressure with the passage of a weather front are less than 1 kPa. Results using Equation 8.4 seem satisfactory in spite of the fact that the specific gas constant for dry air is used.

The question then arises: how typical is the 6.5 K per km lapse rate and the average sea level temperature of 288 K used in the US Standard Atmosphere for places such as Mexico, Canada, or other locations in the world? A brief history of *Standard Atmospheres* will help to answer this question. The original US Standard Atmosphere was developed by the *National Advisory Committee for Aeronautics* (NACA) before World War II. In 1954 a Standard Atmosphere was published by the *International Civil Aviation Committee* (ICAN). For estimating ET, the differences have no importance. The small differences deal with the height of the tropopause and a definition of the acceleration of gravity. Most recent developments deal with extensions to higher altitudes for rocketry. As a result, many committees have been formed, and a long list of acronyms has resulted. Obviously, there have been many attempts at international cooperation and standardization. Most mid-latitude standard atmospheres are the same to about 11 km. Discussion here refers to the US Standard Atmosphere of 1976.

In Climate of the Free Atmosphere (Rex, 1969), a recommendation was made that a separate atmosphere be defined for each 15 degrees of latitude from 15N to 75N for summer and winter conditions. Tentative data for such atmospheres appear in Rex (1969). The atmosphere surrounding the earth varies from the equator to the poles because of mixing that is temperature related. According to Rumney (1968), the tropopause reaches to 15.2 km at the equator and occurs at about 9.1 km at the North Pole. In the supplements to the US Standard Atmosphere, the surface temperature decreases with latitude, and the lapse rate also decreases as movement is northward. Figure 8.1 shows data taken from Rex (1969) and the 1966 Supplement to the U.S. Standard Atmosphere. Probably the US Standard Atmosphere provides reasonable pressure estimates except in equatorial or polar regions.

Table 8.1
Atmospheric pressure comparison, random air pressure measurements compared to estimates using constant lapse rate model, base temperature 288 K, lapse rate 6.5 K or °C per km

location	est. atm. pressure kPa	mea. atm. pressure kPa	error %	remarks
Laramie	77.4	78.6	1.5	Oct. 28, 1986
Logan	84.9	85.0	0.1	summer avg.

Most texts dealing with fluid mechanics, thermodynamics, dynamic meteorology, and hydrology recognize that it is necessary for engineers and others to estimate the average atmospheric pressure for a given location. These references show tables of atmospheric pressure and density based on what is called the US Standard Atmosphere.

8.7.2 *Vapor pressure*

It may be assumed that saturation vapor pressure values shown in the Smithsonian Meteorological Tables (List, 1984) are a standard (Brutsaert, 1982). Data from steam tables (Keenan et al., 1979) may be an even better standard because they are constantly being updated. Many simplified empirical equations have been developed to estimate vapor pressure.

Meteorologists generally work with temperatures below 50 °C. Thermodynamic tables of properties for water vapor have saturation vapor pressure data for temperatures up to about 375 °C. Empirical models of saturation vapor pressure used by meteorologists, therefore, can be simpler because they only need to duplicate values found in steam tables over a narrow range.

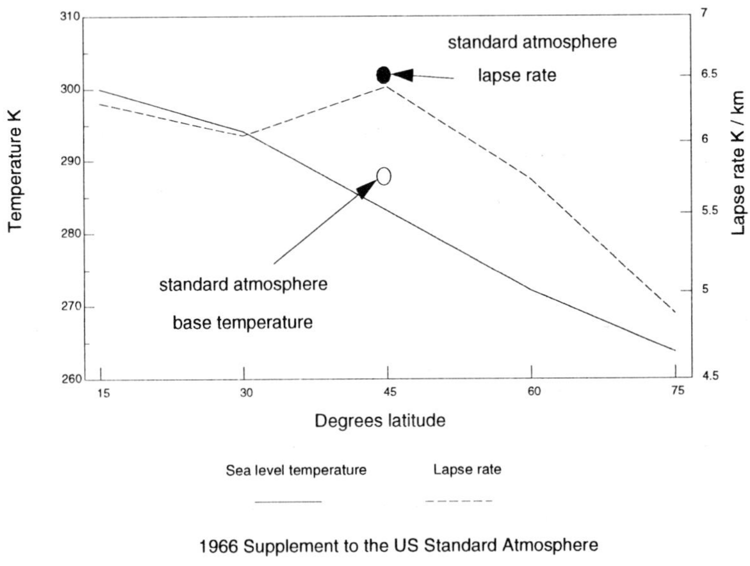

Figure 8.1

Temperature and lapse rate variations

Four methods of estimating saturation vapor pressure are compared with steam table data in Figure 8.2. The equations were selected because each represents a different type of a vapor pressure equation. Equation 8.5 estimates saturation vapor pressure over a wide range of temperatures. Equations 8.6 to 8.8 were developed for meteorological purposes and, therefore, are intended to make satisfactory estimates over a much more limited range.

Weiss (1977) makes comparisons between four other equations for estimating saturation vapor pressure from temperature with similar conclusions. In selecting an equation, care must be observed in making sure that the author's temperature range will not be exceeded and that estimated values are close to standards such as List (1984) or steam tables (Keenan et. al, 1979).

The four equations that follow are written as they are shown in the references, and the units are not consistent nor in accord with SI conventions.

Silver and Nydahl

Equation 8.5 with needed coefficients appears in Silver and Nydahl (1977), and is of a form shown by Wark (1983). The equation is of a classic thermodynamic configuration and is intended to provide accurate estimates over the entire range of temperatures found in steam tables. The origin of the constants is unknown. The authors feel that the number of significant figures shown is excessive.

$$\ln e^o(\text{atmos}) = 70.4346943 - \frac{7362.6981}{T_k} + 0.006952085 T_k - 9.000 \ln T_k \qquad (8.5)$$

Bosen

The Bosen equation was developed for meteorological use and is only intended to provide accurate estimates of saturation vapor pressure over the range of temperatures used in meteorology (Bosen, 1960).

$$e^o = 3.386[(0.00738T + 0.8072)^8 - 0.000019|\,1.8T + 48|\, + 0.001316] \qquad (8.6)$$

Murray

The Murray equation (Murray, 1967) is apparently a variation of those proposed by Tetens (1930) and may be written as

$$e^o = 0.6108 \exp\left[\frac{17.27T}{T + 237.3}\right] \qquad (8.7)$$

Wright

The Wright (1982) equation was determined by multiple correlation using data from the Smithsonian Meteorological Tables.

$$e^o = 0.6105 + 0.0444T + 0.0014347T^2 + 0.00002623T^3 + 0.0000002953T^4 + 0.000000002559T^5 \qquad (8.8)$$

where e^o is *saturation vapor pressure* in kPa unless otherwise noted, T_k is temperature in Kelvin, and T is temperature in °C. These units apply to the four previous equations.

The list of methods of estimating saturation vapor pressure from air temperature is far from complete, but represents methods differing in formulation to illustrate that several satisfactory methods are available.

Estimates of saturation vapor pressure were made using Equations 8.5 through 8.8 for temperatures ranging from near zero to 200 °C. The values were then compared with data from standard steam tables. Figure 8.2 is a comparison of the percent errors resulting from the use of the four methods. When reading about a method of estimating saturation vapor pressure, the data used for calibration or recommendations concerning the range of applicability should be noted. All of the methods provided estimates well within acceptable errors from near zero to about 50°C with errors of about one-half of a percent or less. Estimates made by the method described by Silver and Nydahl (1977) were the closest to standard values being usually under 0.2 % for the entire range of temperatures investigated. The equation described by Silver and Nydahl is intended to estimate vapor pressures for the entire range of values found in common steam tables. The regression equation developed by Wright (1982) surprisingly produced estimates that were satisfactory over the entire range of temperatures with errors less than 1%. The equation of Bosen provided excellent estimates up to about 50°C where the equation produced errors that became increasingly large. The Murray equation produced acceptable estimates up to 90°C. Errors using the Murray equation varied from 0.01% to 0.11% within the desired temperature range. An inspection of Figure 8.2 reveals that Equations 8.5-8.8 will provide satisfactory estimates up to about 50°C.

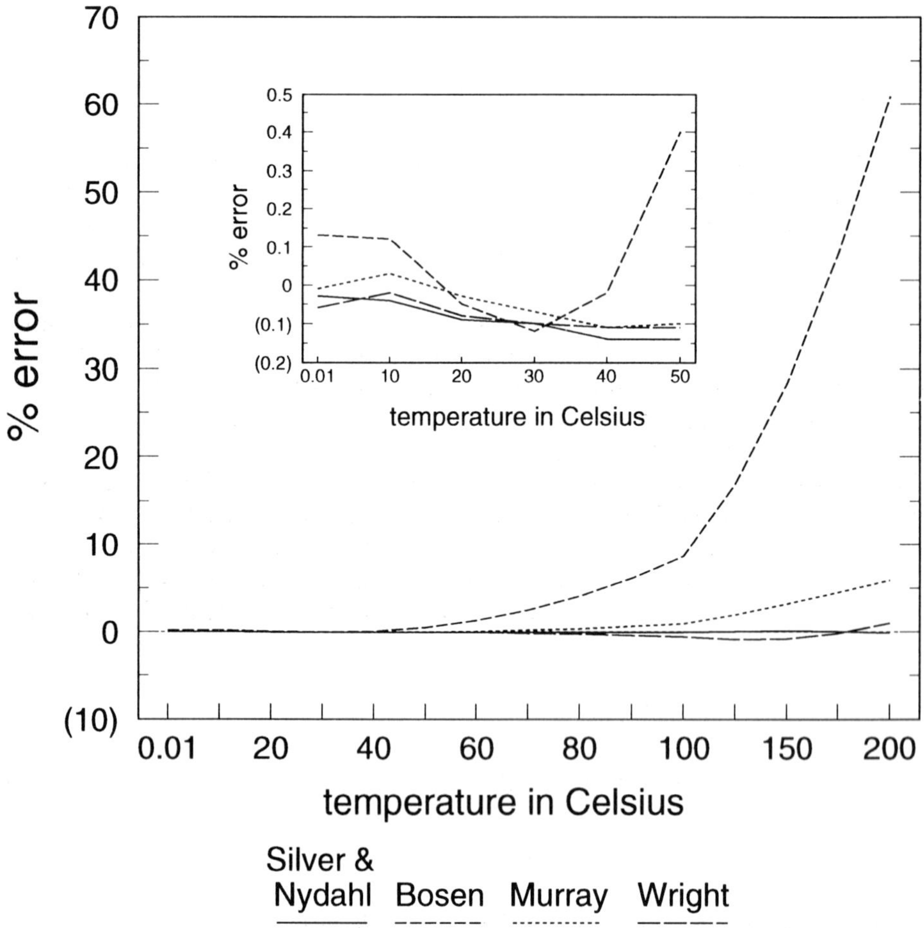

Figure 8.2
Comparison of vapor pressure estimates

Much of the interest in developing alternate models for estimating vapor pressure in the atmosphere apparently was generated by a need to have models that could be used repeatedly with a minimum of computer time as a part of large atmospheric models used for weather prediction. Any one of several methods can be used for estimating ET where computer time is not important. Figure 8.2 shows that all four methods produce satisfactory results at temperatures important in estimating ET.

8.7.3 *Latent heat of vaporization*

Latent heat of vaporization, λ, is used to estimate the so-called psychrometric coefficient, γ, and in converting heat energy to volumes of water. Errors in latent heat of vaporization do not appear to be important in calculating γ but could be significant in converting energy values of ET to volumes of water. λ has the units of kJ kg^{-1}.

Latent heat of vaporization is defined as the quantity of energy required for a phase change from a liquid to a gas, or from liquid water to water vapor, under both a constant temperature and a constant pressure process. In thermodynamic references, this property is known as enthalpy of vaporization (Silver and Nydahl, 1977; Van Wylen and Sonntag, 1985). Values of λ are found in steam tables under columns for *enthalpy*, *h*. Specifically, λ is h_{fg} which is the change in enthalpy required to move from the saturated liquid point to the saturated vapor point within the steam tables. Meteorologists have developed simple procedures for estimating λ from temperatures. At least three variations of simplified methods of estimating latent heat of vaporization are found in the literature.

$$\lambda = 2491 - 2.135T \quad \text{Brunt 1952} \tag{8.9}$$

$$\lambda = 2501 - 2.3601\,T \quad \text{Harrison 1963} \tag{8.10}$$

$$\lambda = 2501 - 2.3668T^{*} \quad \text{Weiss 1977} \tag{8.11}$$

where T is *air temperature* in °C but not designated as to which specific temperature is involved. T^{*} is specifically identified as being the wet bulb temperature in centigrade units. λ values are in kJ kg^{-1}. When λ values are used in the psychrometer equation, the use of the wet bulb temperature is mandatory.

The formulae shown above originally were in the units of cal $gram^{-1}$. To convert to kJ kg^{-1} the relationship between cal and J is required. A value of 1 cal = 4.1868 J was used. This value is not the same in all references and the value used by Weiss is not known. For this reason values of h_{fg} at 0° and the slope of the line will vary between references. Eagleson (1970) used a definition of C_p and shows that the slope of the relationship should have a value of -2.39.

The Brunt method appears in his book on dynamic meteorology (Brunt,

1932). The method credited to Harrison (1963) is a result of using the Goff-Gratch formulation. A numerical check reveals the method duplicates steam table data very well. The methods cited assume a straight line relationship between temperature and ehthalphy when in fact the relationship is not necessarily a straight line. Figure 8.3 is a comparison between the Brunt and Harrison methods and data from the steam tables. Both methods appear to estimate latent heat satisfactorily from 0°C to about 50°C. Both methods appear to estimate latent heat to within 0.5% from 0°C to 70°C. The Harrison method seems to be slightly better than the Brunt method.

Evaporation and transpiration occur at the temperature of the water involved and at a partial pressure equal to the saturation vapor pressure at that temperature. With this assumption, latent heat is only a function of temperature and can be assumed to be the same regardless of atmospheric pressure. Evaporation is the process where saturated water, designated as sub *f*, is moved to saturated gas shown as sub *g*. Latent heat of vaporization is the property enthalpy *h* shown in steam tables. Enthalpy is defined as internal energy plus pressure times specific volume. The $P \cdot v$ term represents work involved with changing density going from a liquid to a gas even though the process is at constant pressure. Data from steam tables show that the pressure-density work is only about 5 to 7% of the work required for evaporation. Most of latent heat is then required to change the internal energy of the water.

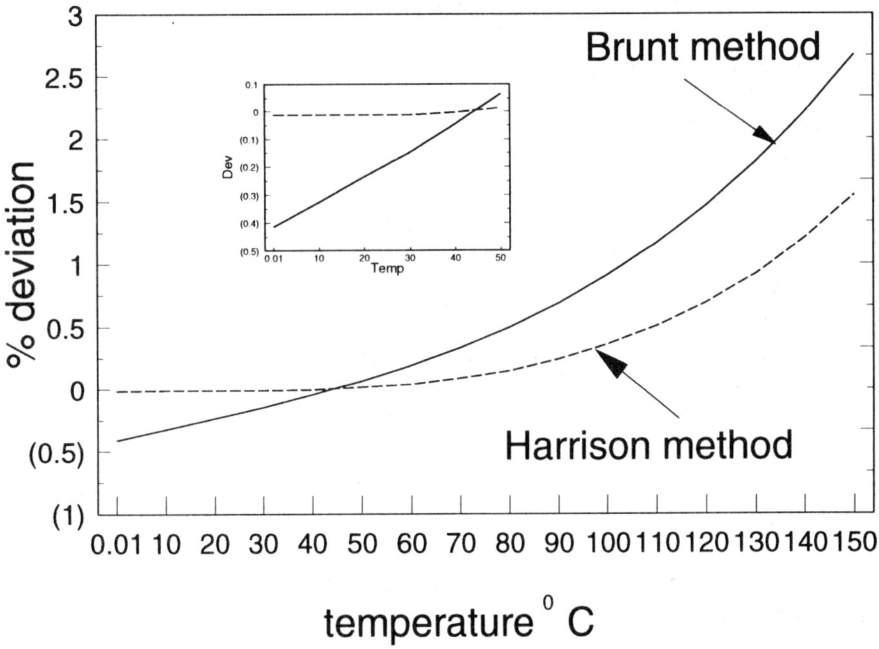

Figure 8.3
Comparison of methods of estimating
latent heat of vaporization, λ

The above equations illustrate inconsistencies in use. Equations 8.9 and 8.10 are based on an unspecified temperature usually assumed to be dry bulb while Equation 8.11 specifies the wet bulb temperature. The question of whether temperature should be a daily average or not is not addressed. The apparent disagreement is really more a matter of interpretation of the appropriate temperature.

Two questions are important in estimating latent heat of vaporization:

1. How accurately does the method estimate latent heat of vaporization as a function of temperature?

2. What temperature should be used in estimating latent heat used in calculating gamma and in converting energy to equivalent volume of water?

The second question concerning temperature is more difficult for estimating ET. Answers will not be perfect, and reasonable assumptions need to be made. Brunt (1952) and Harrison (1963) make the point that the wet bulb temperature should be used in reducing psychrometer data to actual vapor pressure. Derivations of combination theory (Bartholic, et al., 1970) seem to imply that in evaluating gamma, the temperature used should be the temperature at which ET is occurring. In ET and irrigation scheduling literature, there is much discussion concerning plant temperatures. A well-watered crop, such as a reference crop, usually should have a temperature lower than air temperature. A plant under drouth conditions should have a temperature above air temperature. Monteith (1981) discusses surface temperatures to be expected for open water, bare soil, leaves, vegetation, and other conditions. His discussions seems to be based on steady state conditions. Assuming that the shortest time period for using combination methods is twenty-four hours, it is reasonable to use the average daily dry bulb temperature in estimating λ. An alternative is to provide additional weight to the maximum temperature because ET takes place during daytime hours.

Table 8.2
Calculations using latent heat (enthalpy) of vaporization

dry bulb temp °C	relative humidity %	actual vapor pressure kPa	wet bulb temp °C	λ latent heat kJ/kg	γ gamma kPa/°C	delta γ gamma %	ET mm/day	delta ET %
25	00	0.00	6.4	2486	0.0507	-1.8	7.8	-1.8
25	25	0.79	12.4	2472	0.0513	-1.2	7.9	-1.2
25	50	1.59	17.4	2460	0.0518	-0.7	7.9	-0.7
25	75	2.38	21.5	2450	0.0523	-0.3	8.0	-0.3
25	100	3.17	25.0	2242	0.0527	0.0	8.0	0.0

Table 8.2 shows a set of typical calculations for a temperature of 25 degrees C. Relative humidity values of 0, 25, 50, 75, and 100% are assumed. The effect of using dry bulb or wet bulb temperature data is investigated. Calculations assumed a barometric pressure approximating Laramie, Wyoming, and are based on the assumption of λET being 19.5 MJ m^{-2} day^{-1} as an expedient. The assumption of constant ET is questionable because the value of ET is influenced by the latent heat of vaporization. The calculation of the wet bulb temperature is a trial an error procedure based on the psychrometer equation, the vapor pressure relationship, gamma and C_{pm}. A spreadsheet solver was used to calculate T_{wb} which involved all of the relationships

mentioned. When operational estimates of E and ET are made the value of T_{wb} is usually unknown unless the trial and error procedures outlined are utilized. The changes in γ and ET were determined from appropriate values of parameters at 100% humidity

An inspection of Table 8.2 reveals that the changes in the rate of ET and and γ are both less than 2% over the entire range of humidities. The effect of γ on ET estimates is not known at this point. It appears that the practice of basing γ and ET volumes on dry bulb temperature data is reasonable but that further studies of what temperature should be used is warranted.

8.7.4 *Psychrometer coefficient* γ

The psychrometer constant, γ, has been considered to be a constant when it is not really a constant. This is probably because previously calculations were made by hand or with a rotary calculator. In many cases, only a small error was introduced by assuming that γ was constant. The need for simplified methods to calculate γ no longer exists because of the widespread availability of economical personal computers. See Chapter 2 for a discussion of γ.

8.7.5 *Psychrometer derivation*

The pyschrometer equation is based on a simple heat or energy balance which states that the sensible heat defined by the change in air temperature of air entering and leaving a ventilated wet bulb thermometer is equal to the latent energy indicated by the change in water content of the air entering and leaving the psychrometer.

Complex derivations of the psychrometer coefficient are presented by Harrison (1963). The following simplified derivation follows one by Merva (1975).

Energy (lost by wick) = Energy (gained by the air)

$$E_{wick} = E_{air}$$

recognizing that $P \gg e_a$ that $\varepsilon = 0.622$ is defined as the ratio of the mole weights of water vapor to that of air and λ(enthalpy of vaporization) or E(energy) have the units of kJ kg^{-1} and represent energy per unit mass. *Specific humidity*, q, follows:

$$q = \rho_{water}/\rho_{moist\ air} = 0.622\frac{e_a}{P - 0.378\,e_a} \approx 0.622\,\frac{e_a}{P} \qquad (2.18)$$

then E_{wick} is equal to

$$E_{wick} = \frac{0.622\,\lambda}{P}(e^{o}_{wb} - e_a) = \left[\frac{kJ}{kg}\right]\left[\frac{kPa}{kPa}\right] = \frac{kJ}{kg} \qquad (8.12)$$

where the subscript *wb* refers to wet bulb and '*a*' means the surrounding air.

a general text definition of C_p follows:

$$d\mathrm{E} = d\lambda = dh = C_p\, dT \qquad (8.13)$$

then after the integration of Equation 8.13, E_{air} specifically equals:

$$E_a = C_{pm}\,(T_a - T_{wb}) = \left[\frac{kJ}{kg\ {}^\circ C}\right]\left[\frac{{}^\circ C}{1}\right] = \frac{kJ}{kg} \qquad (8.14)$$

equating $E_{wick} = E_{air}$ and $\varepsilon = 0.622$

$$C_{pm}\,(T_a - T_{wb}) = \frac{\varepsilon\,\lambda}{P}(e^{o}_{wb} - e_a)$$

recognizing that $\gamma = \frac{C_{pm}\,P}{\varepsilon\,\lambda}$ and $\Delta T_{wd} = T_a - T_{wb}$

$$\Delta T_{wd}\,\gamma = e^{o}_{wb} - e_a \qquad (2.21)$$

The simple derivation shows that C_p should be for moist air because only moist air both before and after the psychrometer is used in the derivation. An approximation of specific humidity q was used in the derivation.

8.7.6 *specific heat*

C_p is *specific heat at constant pressure* usually used with a value of about 1.004 kJ kg^{-1} K^{-1} at standard sea level conditions in SI units. Different sources show slightly differing values. The value of C_p shown is for dry air. Brutsaert (1982) recommends the following correction to C_p for moist air. Fleagle and Businger (1980) show the same equation but with the constant having the value of 0.9 instead of 0.84.

$$C_{pma} = (1 + 0.84q)C_{pda} \qquad (2.23)$$

where the subscript *pda* refers to dry air, *pma* means moist air, and *q* is *specific humidity* which is dimensionless.

Table 8.3
Estimated specific heat at constant pressure for various temperatures and humidities calculated using Equation 2.23
values are in kJ kg^{-1} K^{-1}

RH	temperature, °C					
%	0	10	20	30	40	50
100	1.0054	1.0074	1.0109	1.0168	1.0267	1.0423
75	1.0049	1.0064	1.0090	1.0135	1.0209	1.0326
50	1.0045	1.0054	1.0072	1.0102	1.0151	1.0229
25	1.0040	1.0045	1.0053	1.0068	1.0093	1.0132
0	1.0035	1.0035	1.0035	1.0035	1.0035	1.0035

It is obvious that C_p changes slowly with both temperature and relative humidity. The use of a generalized constant value of C_p for a reasonable range of conditions encountered seems justified. Values shown are for standard sea level conditions. Based on Table 8.3 a value of C_p for moist air of 1.01 kJ $kg^{-1}K^{-1}$ is justified for typical estimates of ET. Jensen et al. (1990) recommend a value of 1.013 kJ $kg^{-1}K^{-1}$.

When calculating the psychrometer coefficient, C_p for moist air should be used. A constant value reflecting the average conditions likely to prevail can be used. When required ε should then be the ratio of the mole weight of water vapor to the mole weight of dry air. The commonly accepted numerical value of ε is 0.622.

8.8 Sun-earth relationships

The general topic of sun-earth geometric relationships is presented in Chapter 2 Section 2.3.7. Only equations used in Chapter 8 will be repeated along with some materials dealing with units as well as selected explanatory material.

8.8.1 *Sun-earth geometry*

Specific equations needed to estimate various components of solar radiation follow.

Assuming the earth and sun are points and neglecting the refraction of light by the atmosphere the zenith angle, θ_z, may be calculated as follows.

$$\cos\theta_z = \sin L \sin\delta + \cos L \cos\delta \cos\omega_h \qquad (2.28)$$

where θ_z is the *angle between the sun and vertical*, δ is the *declination* of the sun, and L is the *latitude* of a point. All angles are usually in degrees.

The *hour angle* ω_h is the angle between the sun at any time when it is visible and the projection of the sun at solar noon. The following expression, neglecting the diameters of the sun and the earth and atmospheric refraction, is used to estimate ω_h.

$$\cos\omega_h = \sin\left[\frac{\cos\theta_z}{\cos\delta\cos L}\right] - \tan\delta\tan L \qquad (2.29)$$

where all units have been defined previously. The ω_h at sunset reduces to the following because the cosine of the zenith angle is zero at sunrise and sunset.

$$\cos\omega_{hs} = -\tan\delta\tan L \qquad (2.30)$$

All parameters have been defined previously.

8.8.2 *Yearly variations in sun-earth geometry*

For many purposes, such as navigation and cadastral surveying, precise values of sun-earth geometric parameters are essential for an individual day, and differences between years must be recognized in any computation scheme. Because the estimation of evaporation is not an exact science, nearly exact values of parameters, such as declination, are not needed based on the specific year under consideration.

The time required for the earth to orbit around the sun is not the nominal year. The time for a circuit is in excess of a year by about one-fourth of a day. This necessitates the inclusion of an extra day each four years, known as *leap year*. Additional small corrections are also required. For example, this means that a geometric parameter is not exactly the same on all dates of July 1 for various years. These variations cause small errors in parameters, such as clear day solar radiation. These deviations are not of significance in estimating evaporation. However, the user of these parameters needs to be aware of the limitations of data and methods used.

Variations between years are predictable, however, and are published in various places. Examples are *The Astronomical Almanac for the Year 1989* (United States Naval Observatory, 1988a) and *Astronomical Data for the Year 1990* (United States Naval Observatory, 1988b) both published jointly by the US Naval Observatory and the Royal Greenwich Observatory of Great Britain. Both of these publications will be discussed in more detail, and these or similar documents are made routinely available for those who require them. Comparisons between these or similar publications follow.

8.8.3 *Declination*

The North and South Poles are defined by the axis the earth rotates about in the passage of a day. This axis is tilted with respect to the sun. Therefore, the earth's angle with respect to the sun varies approximately ±23.5 degrees as the earth orbits around the sun in the passage of a year. Even though the earth rotates about the sun, many illustrations assume that the earth is stationary and the sun moves.

At any time throughout the year the angle between the sun and the earth is known as the *declination angle*, δ. The sign of δ defines the season. Positive values of delta are defined as summer in the Northern Hemisphere and winter in the Southern Hemisphere. The reverse is true for negative values of δ. The tropical areas of the world lie within latitudes defined by the positive and negative values of δ. Polar regions of the world are those areas where the latitude and δ lead to complete darkness in any twenty-four hour period of the year.

8.8.4 *Cooper model* δ

The procedure attributed to Cooper (1969) is very easy to use in both spreadsheet calculations and in programming with conventional computer languages and appears widely in various publications. See Section 2.3.9 for additional information. The algorithm follows:

$$\delta = 23.5 \sin\left[\frac{360(D + 284)}{365}\right] \qquad (2.34)$$

where δ is in degrees, and D is the *calendar day number*. The argument of the sine function is also in degrees.

8.8.5 *Burman and Jacquot model* δ

This model is described in detail in Section 2.3.10. The following periodic equation is used. Slight errors will occur from year to year because of yearly variations in dates and leap years. These errors are so small that reasonable estimates of evaporation should be produced by this method. For comparative purposes the following equation has an additional term compared to Equation 2.35.

$$\delta = 0.3931 + 23.2577\cos(0.9861\,D - 170.7) + 0.3906\cos(1.9154\,D - 174.4)$$

$$+0.176\cos(3.01721\,D - 153.5) \qquad (2.35)$$

where *delta* is in degrees, and D is the *calendar day number* of the year.

8.8.6 *Spencer model* δ

A periodic analysis of δ was made by Spencer (1971). This involves the use of a parameter Γ which is determined as follows.

$$\Gamma = 2\pi\left[\frac{D-1}{365}\right] \qquad (8.15)$$

where D is the *calendar day number* of the year, and Γ is in radians.

The fundamental equation now follows.

$$\delta = \left[\frac{180}{\pi}\right][0.006918 - 0.3999\cos(\Gamma) + 0.070257\sin(\Gamma)$$

$$-0.006758\cos(2\Gamma) + 0.000907\sin(2\Gamma)$$

$$-0.002967\cos(3\Gamma) + 0.00148\sin(3\Gamma)] \qquad (8.16)$$

where δ is in degrees.

8.8.7 *Stine and Harrigan model* δ

Stine and Harrigan (1985) state that for many solar design purposes an approximation of δ to the nearest degree is adequate. The same is true of estimating E or ET. It is not known if the book by Stine and Harrigan is the primary reference. Their trigonometric equation follows:

$$\sin \delta = 0.39795 \cos[0.98563(D - 173)] \tag{8.17}$$

where δ is in degrees and D is the calendar day number.

8.8.8 *Day-length*

One of the two data requirements for the Blaney-Criddle method of estimating evapotranspiration is the percent of daylight hours of a year for the period in question. FAO-24 version of the Blaney-Criddle method requires that the percentage be based on a daily basis, but the earlier version utilized the parameter p which is the monthly percent of theoretical yearly sunlight. Blaney and Criddle (1962) originally utilized tabular data from the US Naval Observatory (1946) and the Smithsonian Meteorological Tables (List, 1984). The percentage p is, therefore, a quantity that is only related to latitude and date.

The following development neglects the diameters of the sun and earth and also neglects the refraction of light rays by the earth's atmosphere. Both refraction and the diameter of the sun are included in astronomical information appearing in solar ephemeris publications. Ignoring the diameters of the sun and earth as well as refraction, the following expression may be used to estimate day length. The topic is discussed in Section 2.3.13.

$$N = \omega_{hs}\left[\frac{2 \cdot 24}{360}\right] \tag{2.37}$$

where N represents the number of hours of theoretical sunlight with the limitations mentioned above.

8.8.9 *Percent daytime*

The early versions of the Blaney-Criddle method (Blaney and Criddle, 1962) require numerical values of the monthly percent daytime on an annual basis. Doorenbos and Pruitt (1977) specify percent daytime on a daily value related to the annual total.

The preferred method of calculation would be first to calculate daytime hours for each day of the year. The desired percentage can then easily be determined by dividing the sum of daytime hours for the period in question by the annual sum which is then multiplied by 100.

An alternative method would be to calculate the day length for the midpoint of each month. Then the annual total number of daytime hours can be developed by taking the sum of monthly hours determined by multiplying the number of days in each month by the mid-month value of daytime hours.

Monthly percent values of daytime hours can then be calculated by using the product of 100 times the monthly total divided by the yearly total hours of daytime. This topic is also discussed in Section 2.3.14.

8.8.10 *Earth-sun distance correction*

The orbit of the earth about the sun is slightly elliptical in shape, thereby causing the distance between the two bodies to vary slightly. Heat transfer by radiation is known to vary inversely with the distance squared. The distance between the sun and the earth is very large, and for this reason a correction based upon the square of the relative distance between the earth and sun is applied to extraterrestrial radiation estimates. Iqbal (1983) shows 1.496 times 10^5 km as being the average distance between the sun and the earth. The correction has the form of $\left[\frac{\bar{d}}{d}\right]^2$. Distance d is the distance between the earth and sun, and $\bar{d}$ is the average distance between the two bodies.

The functional relationship is known to be periodic. For this reason, empirical formulae using sines or cosines are used to estimate the correction. Two algorithms for estimating the correction needed because of the slightly elliptical distance between the sun and the earth are discussed, and the results of comparing the estimates with standard values are discussed.

8.8.11 *Duffie-Beckman model*

The expression appearing in Duffie and Beckman (1980) is:

$$\left[\frac{\bar{d}}{d}\right]^2 = 1 + 0.034\cos\left[\frac{360D}{365.25}\right] \tag{8.18}$$

where the distance correction $\left[\frac{\bar{d}}{d}\right]^2$ is dimensionless, and D is the *calendar day number.*

8.8.12 *Spencer distance correction*

A more complex Fourier analysis of the sun-earth distance correction resulted in the following periodic series (Spencer, 1971).

$$\Gamma = \frac{2\pi(D-1)}{365} \tag{8.19}$$

$$\left[\frac{\bar{r}}{r}\right]^2 = 1.00011 + 0.034221\cos\Gamma + 0.00128\sin\Gamma + 0.000719\cos 2\Gamma + 0.000077\sin 2\Gamma \tag{8.20}$$

where D is the *calendar day number* of the year.

8.8.13 *Extraterrestrial solar radiation*

Extraterrestrial radiation, R_a, is the solar radiation received on a horizontal surface above the earth's atmosphere and is very useful in predicting ground level solar radiation. Extraterrestrial radiation, R_a , is only a function of latitude and date if radiation from the sun perpendicular to the sun's rays is assumed to be constant. Because the earth's axis is tilted and the distance from the sun to the earth varies with season, the geometry involved is complex but merely represents an integration of the functional relationship between the zenith angle of the sun with an empirical distance correction. This procedure ignores the diameters of the sun and earth and refraction of light rays because of the atmosphere.

$$R_a = \frac{SC}{\pi}\left[\frac{\bar{d}}{d}\right]^2\left[\omega_{hs}\left\{\frac{180}{\pi}\right\} \sin L \sin\delta + \cos L \cos\delta \cos\omega_{hs}\right] \qquad (2.41)$$

where SC is the *solar constant* in MJ m^{-2} day^{-2}, and R_a is *extraterrestrial solar radiation* in MJ m^{-2} day^{-1}.

8.8.14 *Clear day solar radiation*

Clear day solar radiation, R_{so}, refers to the radiation received at any place on the earth when the effects of the atmosphere are included, and the effects of clouds are excluded. For this reason, R_{so} is often known as *cloudless day radiation.* In the estimation of net radiation for evaporation the ratio of actual to clear day solar radiation is used as an input parameter. It is relatively easy to measure solar radiation using present day microprocessor controlled instrumentation. The operational measurement of net radiation is, however, usually much more difficult than the measurement of solar radiation.

Values of clear day solar radiation may easily be determined by plotting an extensive series of solar radiation on a time scale. Clear day solar radiation is then merely the upper envelope of the measured values. Estimates of clear day solar radiation are needed until a measured curve can be developed. Then a suitable trigonometric function useful in prediction models can be determined.

Much of the information dealing with the estimation of clear day radiation comes from the heating, ventilating, and solar collector industries. Energy symbolized by clear day radiation is the maximum possible for use in a solar collector, or the maximum value of energy that would need to be removed in a building ventilating system. In a solar collector, clear day radiation would be the maximum energy available at any time. Models for the estimation of actual solar radiation are available, usually as a function of cloud cover.

Including the effects of the atmosphere is indeed a complex issue. The sun's radiation is attenuated by normal constituents of the air such as nitrogen or water vapor and colloidal particles known aerosols. Dust or smoke particles are examples of aerosols. Available models range from rather simple to very complex models that require large digital computer programs. Bird and Hulstrom (1981) describe the use of the *SOLRAN* in estimating clear day solar radiation, R_{so}.

8.9 Comparisons with standard values

Data from several publications have become accepted as being standard by professionals who estimate evaporation and evapotranspiration. These include the Smithsonian Meteorological Tables (List, 1984) for many parameters and various publications of Budyko for radiation phenomena. Many additional sources that may be regarded as standard exist as well. The additional standards may include various ephemeris publications and various publications of ASHRAE, the American Society of Heating, Refrigerating and Air Conditioning Engineers, Inc.

The Smithsonian Meteorological Tables (List, 1984) were first published in 1951, according to a copy purchased in the spring of 1990. The first reprint with corrections was issued in 1958. The current volume is shown as the 5th reprint, dated 1984. The publisher notes that revisions of some Sections are needed, but that demand for the volume is active and that a qualified source of such a revision cannot be found. Even though the Smithsonian Meteorological Tables (List, 1984) show a publication date of 1984, most of the calculations are based on data for the year 1950.

Wherever possible, standards that are widely accepted have been used for comparisons. In addition, information from other sources not commonly used by professionals in estimating evaporation have been included.

8.9.1 *Statistical and numerical comparisons*

Simple numerical tests, such as the maximum error and the time of maximum error, are used extensively in comparing the performance of algorithms with standard values. In addition, the *standard error of the estimate*, SEE, was used. The SEE may be calculated as follows:

$$SEE = \sqrt{\frac{[x_s - x_c]^2}{N-1}} \tag{8.21}$$

where x_s is the standard value of a parameter, x_c is the estimated value of the parameter, and N is the number of comparisons. The Irrigation Water Requirements Committee of the American Society of Civil Engineers (Jensen et al., 1990) use the SEE in comparing the performance of various methods of estimating evapotranspiration.

8.9.2 *Declination comparisons*

Publications such as a solar ephemeris or the publication *Astronomical Data for the Year 1990*, jointly published by the US Naval Observatory and the Royal Greenwich Observatory (1988b) show estimates of daily declination for a specific year. List (1984) shows values for the year 1950. If an algorithm will predict δ for one year, it should have the capability of making at least reasonable estimates for any other year. Certainly both the information from List (1984) and that for other selected years would both make excellent reference data.

The declination estimates made by the US Naval Observatory and the Royal Greenwich Observatory represent a specific year rather than an average. The 365 daily values appearing in the publication *Astronomical Data for the Year 1990* were compared with estimates of δ made by the models of Burman and Jacquot, Spencer (1971), Cooper (1969), and Stine and Harrigan (1985). Comparisons for the Stine and Harrigan model were made using a spreadsheet because the PASCAL compiler did not have an inverse sine function. All other comparisons were made using a custom PASCAL computer program.

The model attributed to Burman and Jacquot was developed after the other models predicted δ values with errors up to 1.28°. Other comparisons show that such an error was not of practical significance although such an error seems excessive.

The maximum error in degrees, the date of the maximum error, and the *standard error of the estimate,* SEE, were determined for each variation. The Burman-Jacquot model was evaluated using 1, 2 and 3 terms of the series. The method of Spencer was evaluated for 5 and 7 terms of the series. The Cooper and Stine and Harrigan methods have only a single term. The results follow.

Table 8.4
Errors in estimating the declination of the sun
numerical values are in degrees

model	terms	standard error	max error	error date	1990 decl
Burman-Jacquot	1	0.301	-0.57	Dec 31	-23.13
Burman-Jacquot	2	0.125	0.21	Nov 7	-16.05
Burman-Jacquot	3	0.041	0.15	Nov 7	-16.05
Spencer	5	0.155	0.30	Feb 20	-11.08
Spencer	7	0.092	-0.21	Oct 15	-18.37
Cooper	1	0.567	1.28	Oct 11	-6.83
Stine and Harrigan	1	0.463	0.79	Oct 2	-3.38

The Burman-Jacquot method was developed using data from the US Naval Observatory and Royal Greenwich Observatory, Astronomical Data, 1990.

An inspection of Table 8.4 reveals that all models investigated will probably estimate the declination of the sun, δ, to the precision needed for the estimation of E or ET. No particular pattern appears for the dates of occurrence maximum errors. All models are periodic, and as a result involve sines, cosines, and often more than one term. It is obvious that the use of additional terms improves the precision of declination estimates.

The model of Cooper is the simplest to use and appears to be the most widespread in its use. The maximum error using the Cooper model is about 1.3°, but this occurs in October which is not part of the active growing season in many parts of the northern hemisphere. Errors using the model of Cooper are much less for the growing season. Recommendations are to use the Cooper model because of simplicity and acceptance, or two terms of the Burman-Jacquot model for improved precision. Two terms of the Burman-Jacquot model are simple to use. The models of Spencer and Stine and Harrigan are more complex to use, and some compilers and spreadsheets may not have the required trigonometric functions.

8.9.3 *Earth-sun distance comparisons*

A numerical correction because of the slightly elliptical traverse of the earth about the sun is required to estimate correctly extraterrestrial solar radiation. The algorithms of Duffie and Beckman (1980), as well as that of Spencer (1971), were compared with tabular values of the relative sun-earth distance shown in List (1984) which were based on 1950 data, and 1989 estimates made by the US Naval Observatory and Royal Greenwich Observatory (1988a).

This presented the opportunity to test the statement made in this volume and many others that very slight variations in parameters occur between dates in different years. This is because of the fact that the time needed for the earth to make one rotation about the sun is not exactly 365 days and leads to the use of leap years and occasional corrections. These variations are said to be too small for practical significance in estimating evaporation.

The tables of relative distances contained in the Smithsonian Meteorological Tables have only eight data points per month. For this reason, the seventeenth of each month was selected as being near the center of each month, and the models were compared to both the 1950 and 1989 data points.

Table 8.5
Comparisons of the sun-earth distance corrections
for different years and two prediction models

comparison	SEE	max error
1950 and 1989 ephemeris estimates	0.01214	0.047%
Duffie-Beckman model vs. 1950 ephemeris	0.00149	0.241%
Duffie-Beckman model vs. 1989 U. S. Navy	0.00157	0.245%
Spencer model vs. 1950 ephemeris	0.00061	0.077%
Spencer model vs. 1989 U. S. Navy	0.00061	0.092%

An inspection of the previous table reveals that differences between a sun-earth distance based on the estimates shown by List (1984) for the year 1950 and estimates shown by the US Naval Observatory and the Royal Greenwich Observatory for the year 1989 are minimal. The SEE was 0.01214, and the maximum error was only 0.05%. This indicates that the variations between years in the sun-earth distance are insignificant when estimating evaporation.

Differences in the earth-sun correction based upon the two models considered and corrections using standard estimates are also very small. The Spencer model appeared to yield corrections fitting standard data slightly better that those using the Duffie-Beckman model. Corrections based upon either model appear to be very near corrections based upon standard references, and the only criteria for the selection of a model should be its ease of use. Based on convenience, the model of Duffie-Beckman model is recommended.

8.9.4 *Day-length comparisons*

Day-length values were required for the Thornthwaite method of estimating evapotranspiration. The Thornthwaite method is not widely used at the present time except, perhaps, as a tool in geography. Day-length is not directly used as an input parameter in methodology described in this volume. For the Blaney-Criddle method, day length is used indirectly in the determination of p the percent of daytime hours on a seasonal basis.

Two sources were considered in evaluating day-length estimates. The Smithsonian Meteorological Tables (List, 1984) have tabular estimates of day length. The publication *Astronomical Data for the Year 1990* (US Naval Observatory, 1988b) has daily estimates of day length for the year 1990.

Day length was calculated using Equation 2.37 and then compared with predictions shown in *Astronomical Data for the Year 1990* (US Naval Observatory, 1988b). Errors are summarized in Table 8.6.

Table 8.6
Day length percent errors between calculated and 1990 US and Royal Observatories for various dates and latitudes

date lat.	Jan 14	Feb 15	Mar 15	April 16	May 14	June 15	July 17	Aug 14	Sept 15	Oct 17	Nov 14	Dec 16	Avg
40	1.8	2.4	1.4	1.2	1.2	1.3	1.1	0.9	0.8	1.3	1.5	1.6	1.4
20	1.1	1.4	1.1	0.8	1.0	1.0	0.9	0.8	0.9	1.1	1.2	1.0	1.0
0	1.1	1.0	0.8	0.8	0.4	1.0	1.0	1.0	1.0	1.0	1.0	1.0	0.9
-20	1.1	0.6	3.6	1.0	1.1	1.1	1.2	1.1	1.1	1.0	1.0	1.2	1.3
-40	1.2	0.5	0.9	1.2	1.6	1.6	1.7	1.8	1.6	1.2	1.2	1.4	1.3
Avg	1.2	1.2	1.6	1.0	1.0	1.2	1.2	1.1	1.1	1.1	1.2	1.2	1.2

The estimates were always less than the predictions because the predictions shown in *Astronomical Data for the Year 1990* (US Naval Observatory, 1998b) include corrections for the radius of the sun and refraction caused by the sun's rays striking the atmosphere.

8.9.5 *Percent daytime comparisons*

The original tables of annual daytime percentage appearing in publications by Blaney and Criddle (1962) have become regarded as being standard. At least two changes in their use have occurred.

The modification of FAO-24 by Doorenbos and Pruitt (1977) uses a percent based on a daily value instead of a month. The table of Doorenbos and Pruitt (1977) utilizes only two significant figures, where the original tables by Blaney and Criddle (1962) show three significant figures for winter months and four significant figures for summer months. Probably the use of three significant figures is justified. The use of only two significant figures tends to introduce additional errors that are not necessary.

Many tables of the annual percent daytime show equal values for both the Northern and Southern Hemispheres with values being merely offset by a period of six months in the alternative hemisphere. The tables of Blaney and Criddle (1962) show slightly different values when applying the six month offset. For example at a latitude of 30°, the February value in the Northern Hemisphere should be compared with the August value in the Southern Hemisphere. The value at 30°N is 6.86 in February, and the value at 30°S for August is 7.80. The difference is largely explained by the difference in the length of the months. The tables showing equal values with a six months' offset are probably within the accuracy of the Blaney-Criddle method but introduce additional variations in results.

The percent daytime was estimated using Equations 2.38 and 2.39 for a point near the middle of each month for latitudes ranging from 60°N to 50°S. The hours of daylight for each month were determined by multiplying the mid-month value by the number of days in each month. The percent daytime for each month was simply determined by dividing by the yearly total and multiplying by 100. The following table shows differences in estimating the percent daytime.

Table 8.7
Differences in the percent daytime
Blaney and Criddle (1962) and estimated
values are percentages

Lat	Jan	Feb	Mar	April	May	June	July	Aug	Sept	Oct	Nov	Dec	Avg
60	2.8	1.4	-1.0	11.4	-2.9	-3.4	-3.3	-2.8	-1.3	0.4	2.8	4.4	0.70
50	0.8	0.4	0.0	8.7	-1.5	-2.0	-2.3	-1.9	-1.2	-0.4	0.6	0.6	0.16
40	0.0	0.3	-0.2	6.0	-1.3	-1.4	-0.4	-1.2	-0.4	0.1	0.3	0.1	0.15
30	0.1	0.2	0.1	4.5	-0.8	-0.9	-1.1	-0.9	-0.4	-0.1	0.0	-0.3	0.03
20	-0.2	0.0	-2.7	2.7	-0.5	-0.5	-0.5	-0.6	-0.2	0.1	0.1	2.8	0.04
10	-0.3	-1.0	-0.2	2.1	-0.2	-0.7	-0.5	0.0	-0.1	-0.6	-0.2	-0.1	-0.15
0	0.0	0.0	0.0	0.0	0.0	0.0	0.0	0.0	-0.4	0.0	0.0	0.0	-0.05
-10	0.0	-0.1	-0.1	-1.2	0.3	0.4	0.1	0.4	0.1	0.1	-0.1	0.1	-0.02
-20	0.1	0.1	0.1	-2.9	0.4	0.5	0.5	0.8	0.1	0.3	-0.1	0.0	-0.01
-30	0.1	0.1	0.2	-4.6	0.5	1.0	1.1	1.4	0.3	0.2	0.0	-0.2	0.01
-40	0.1	0.3	0.4	-6.9	1.1	1.4	1.6	1.8	0.5	0.3	-0.2	-0.2	0.02
-50	0.2	0.4	0.4	-10.3	2.1	2.5	3.2	2.9	0.6	0.2	-0.4	-0.2	0.13
Avg	0.30	0.18	-0.25	0.80	-0.25	-0.26	-0.15	-0.01	-0.20	0.04	0.22	0.58	0.08

Table 8.7 is a summary of errors when the estimates were compared with the values shown by Blaney and Criddle (1962). Most of the estimates are within 1% of the original tabular values. Excessive errors are found only at extreme latitudes or in the month of April. The errors in April cannot be explained by the authors.

8.9.6 *Extraterrestrial radiation comparisons*

All values of extraterrestrial solar radiation appearing in standard references appear to have been calculated using the same theory. The radius of the sun is neglected, and atmospheric refraction is nonexistent by definition. Estimates using Equations 2.40 and 2.41 were compared to values shown in List (1984), and, as expected, were very close to the estimates. Differences encountered appeared to be a matter of significant figures and round-off. Tabular values of errors are not shown because all references known to the authors use the same theory to construct tables and graphs.

REFERENCES

Abtew, W., J.M. Gregory and J. Borrelli, 1989, Wind profile: estimation of displacement height and aerodynamic roughness, Trans. ASAE **32**(2):521-527.

Alexander, E.B., 1989, Bulk density equations for southern Alaska soils, Can. J. Soil Sci. **69**:177-180.

Allen, R.G., 1986, A Penman for all seasons, J. Irrigation and Drainage, ASCE **112**(4):348-368.

Allen, R. G., and W. O. Pruitt, 1986, Rational use of the Blaney-Criddle formula. J. Irrigation and Drainage, ASCE **112**(IR2):139-155.

Allen, R.G., M.E. Jensen, J.L. Wright and R.D. Burman, 1989, Operational estimates of reference evapotranspiration, Agronomy J. **81**(4):650-662.

American Society of Agronomy--Soil Science Division, 1986, Methods of Soil Analysis, Part 1, Physical and Mineralogical Methods, Agronomy Monograph No. 9 (2nd Ed.)

Andersson, L., and R.J. Harding, 1991, Soil moisture deficit simulations with models of varying complexity for forests and grassland sites in Sweden and the U.K., Water Resource Management **5**:23-46.

Angstrom, A., 1924, Solar and terrestrial radiation, Quart. J. Royal Meteorological Soc. **50**:121-125.

ASHRAE Handbook, 1987, Chapter 47: Solar Energy Utilization, American Society of Heating, Refrigerating and Air Conditioning Engineers, Inc., 1791 Tullie Circle, NE, Atlanta, GA., pgs. 47.1-48.28.

Baier, W., and G.W. Robertson, 1966, A new versatile soil moisture budget, Can. J. Plant Sci. **46**:299-315.

Barfield, B.J., and J.F Gerber, eds., 1979, Modification of the Aerial Environment of Plants, Monograph No. 2, ASAE, St. Joseph, MI., 538 pgs.

Bartholic, J.F., L.N. Namken and C.L. Wiegand, 1970, Combination equation used to calculate evaporation and potential evaporation, Agricultural Research Service, USDA 41-170, 14 pgs.

Becker, C.F., and J.S. Boyd, 1957, Solar radiation availability on surfaces in the United States as affected by season, orientation, latitude, altitude and cloudiness, Solar Energy **1**(1):13-21.

Bird R.E., and R.L. Hulstrom, 1981, Review, evaluation and improvement of direct irradiance models, Trans. ASME **103**:182-192.

Black, J.N., 1956, The distribution of solar radiation over the earth's surface, Arch. Meteorology Geophys. Bioklim. (7):165-189.

Blad, B.L., and R. Lemeur, 1979, Miscellaneous techniques for alleviating heat and moisture stress, in: Modification of the Aerial Environment of Plants, Monograph No. 2, ASAE, St. Joseph, MI., 538 pgs.

Blaney, H. F., and K.V. Morin, 1942, Evaporation and consumptive use of water empirical formulas, Amer. Geophysic. Union Trans. Pt. 1 (August):76-83.

Blaney, H.F., and W. D. Criddle, 1945, Determining water requirements in irrigated areas from climatological data, (processed), 17 pgs.

Blaney, H. F., and W. D. Criddle, 1950, Determining water requirements in irrigated areas from climatological and irrigation data, SCS-TP-96, USDA (processed), 48 pgs.

Blaney, H.F., and W.D. Criddle, 1962, Determining Consumptive Use and Irrigation Water Requirements, Tech. Bull. 1275, Agricultural Research Service, USDA, 59 pgs.

Bond, J.J., and W.O. Willis, 1970, Soil water evaporation: First stage drying as influenced by surface residue and evaporation potential, Proc. Soil Soc. Amer. **34**:924-928.

Boonyatharokol, W., and W.R. Walker, 1979, Evapotranspiration under depleting soil moisture, J. Irrigation and Drainage, ASCE **105**(IR4):391-402.

Borrelli, J., L.O. Pochop, W.R. Kneebone, I.L. Pepper, R.E.Danielson, W.E. Hart and V. B. Youngner, 1981, Blaney-Criddle coefficients for western turf grass, J. Irrigation and Drainage, ASCE **107**:331-341.

Borrelli, J., K.J. Fornstrom, D.J. Brosz and R.W. VanKlavern, 1986, Irrigation scheduling, the checkbook method for Wyoming, Univ. of Wyoming, Agric. Exp. Station, B-806, 20 pgs. mimeo.

Bosen, J.F., 1960, A formula for approximation of the saturation vapor pressure over water, Monthly Weather Review **88**(8):275-276.

Brady, R. A., S. M. Goltz, W. L. Powers, and E. T Kanemasu, 1974, Relation of soil water potential to stomatal resistance of soybean, Agronomy J. **67**, January-February 1975, pgs. 97-99.

Brinker, Russell C., and Paul R. Wolf, 1977, Elementary Surveying, Harper & Row, Publishers, New York, NY, 568 pgs.

Brunt, D., 1932, Notes on radiation in the atmosphere: Quart. J. Royal Meteorological Soc. **58**:389-420.

Brunt, D., 1952, Physical & Dynamical Meteorology, 2nd Ed., Cambridge University Press, 427 pgs.

Brutsaert, W., 1975, On a derivable formula for long wave radiation from clear skies, Water Resources Research **11**:742-744.

Brutsaert, W., 1982, Evaporation Into the Atmosphere, D. Reidel Publishing Co., Dordrecht, Holland, 299 pgs.

Budyko, M.I., ed, 1963, Guide to the Atlas of the Heat Balance of the Earth, (translated from Russian) U.S. Dept. of Commerce, WB/T 106, 27 pgs.

Burman, R.D., 1976, Intercontinental comparison of evaporation estimates, J. Irrigation and Drainage ASCE (IR1):109-118.

Burman, R.D., 1985, Comparison of monthly estimates of grass and alfalfa ET, 1985, Proc. Advances in Evapotranspiration Symposium, ASAE, Chicago, IL., December.

Burman, R.D., and J.R. Borrelli, 1984, Water use across a mountain meadow valley, Proc. Second Intermountain Meadow Symposium, Gunnison, CO., July 11-13, Colo. State Univ. & Univ. of Wyoming.

Burman, R.D., J.L. Wright and M.E. Jensen, 1975, Changes in climate and estimated evaporation across a large irrigated area in Idaho, Trans. ASAE **18**(6):1089-1091, 1093.

Burman, R.D., P.R. Nixon, J.L. Wright and W.O. Pruitt, 1980, Water requirements, in: Design and Operation of Farm Irrigation Systems, M.E. Jensen, ed., ASAE, St. Joseph, MI.

Burman, R.D., R.H. Cuenca and Albert Weiss, 1983, Techniques for estimating evapotranspiration in: Advances in Irrigation, Vol. 2, D. Hillel, ed., Academic Press, New York, 335 pgs.

Burman, R.D., M.E. Jensen and R.G. Allen, 1987, Thermodynamic factors in evapotranspiration, in: James, L.G. and M.J. English (eds.), Proc. Irrigation and Drainage Spec. Conf., Portland, OR., July, pgs. 28-30.

Calder, I.R., R.J. Harding and P.T.W. Rosier, 1983, An objective assessment of soil-moisture deficit models, Hydrology **60:**329-355.

Campbell, G.S., 1977, An Introduction to Environmental Biophysics, Springer-Verlag, New York, 159 pgs.

Campbell, G.S., and D.J. Mulla, 1990, Measurement of soil water content and potential, Chapter in: Irrigation of Agricultural Crops, No. 30, Amer. Soc. of Agronomy, Madison, WI., pgs. 127-142.

Chaudhary, M.R., R. Khera and C.J. Singh, 1991,Tilliage and irrigation effects on root growth, soil water depletion and yield of wheat following rice, J. of Agricultural Science, **116**, 9-14.

Cheness, J. L., L.A. Harper and T.A. Howell, 1979, Sprinkling for heat stress reduction, in: Modification of the Aerial Environment of Plants, Monograph No. 2, ASAE, St. Joseph, MI., 538 pgs.

Christiansen, J.E., 1968, Pan evaporation and evapotranspiration from climatic data, J. Irrigation and Drainage, ASAE **94**:243-265.

Christiansen, J.E., and G. H. Hargreaves, 1969, Irrigation requirements from evaporation, Trans. Irrigation and Drainage, ASAE **III** **23**:569-596.

CH2M HILL, 1986, A quick reference guide for selecting crop water production functions, A report prepared for the U.S. Bureau of Reclamation, Denver Office, by CH2M HILL, Sacramento, CA.

Constable, G.A., L.J. Rochester and A.S. Hodgson, 1990, Irrigation science **11**:137-142.

Cooper, P.I., 1969, The absorption of solar radiation in solar stills, Solar Energy **12**(3):333-336.

Cuenca, R.H., 1989, Irrigation System Design: an Engineering Approach, Prentice-Hall, Inc., Englewood Cliffs, NJ.

Cuenca, R. H., and M. T. Nicholson, 1982, Application of the Penman equation wind function, J. Irrigation and Drainage, ASCE IR1:13-23.

Dastane, N.G., 1974, Effective Rainfall in Irrigated Agriculture, Food and Agriculture Organization United Nations, Irrigation and Drainage Paper 25, Rome, 62 pgs.

Daugherty, R.L., and J.B. Franzini, 1977, Fluid Mechanics With Engineering Applications, 7th revision, McGraw-Hill Book Company, New York.

Davis, R.E., and F.S. Foote, 1953, Surveying Theory and Practice, McGraw-Hill Book Company, New York, 1021 pgs.

de Jong, R., and D.R Cameron, 1979, Computer simulation model for predicting soil water content profiles, Soil Sci. **128**(1):41-48.

De Wit, C.T., 1958, Transpiration and crop yields, Versi Landbook Onderz No. 54, Wagenengen, The Netherlands.

deBruin, H.A.R., 1978, A simple method for shallow lake evaporation, J. Applied Meteorology **17**:1132-1134.

Denmead, O.T., and R.H. Shaw, 1962, Availability of soil water to plants as affected by soil moisture content and meteorological conditions, Agronomy J. **54**:385-390.

Doorenbos, J., and W.O. Pruitt, 1977, Guidelines for Predicting Crop Water Requirements, Food and Agriculture Organization United Nations, FAO Irrigation and Drainage Paper 24, 2nd Ed., Rome, 156 pgs.

Doorenbos, J., and A.H. Kassam, 1979, Yield Response to Water, Food and Agriculture Organization United Nations, FAO-33, Rome, 193 pgs.

Duffie J.A., and W.A. Beckman, 1980, Solar Engineering of Thermal Processes, John Wiley and Sons, New York.

Dunin, F. X., A. R. Aston, and W. Reyenga, 1978, Evaporation from a *Themeda* grassland, J. Applied Ecology, *15*:847-858.

Eagleson, P.S., 1970, Dynamic Hydrology, McGraw-Hill Book Co. New York, 462 pgs.

Environmental Protection Agency, 1984, The Hydrologic Evaluation of Landfill Performance (HELP) Model, Office of Solid Waste and Emergency Response, United States, EPA, Washington, DC.

Environmental Science Services Administration, 1966, United States Standard Atmosphere Supplements, 1966, prepared by the Environmental Science Services Administration, National Aeronautics and Space Administration and the United States Air Force, Washington, DC.

Feddes, R. A., Crop Factors in Relation to Makkink Reference-Crop Evapotranspiration, in: Evaporation and Weather, Proceedings and Information No. 39, Technical Meeting 44, Ede. The Netherlands,s 25 March 1987, The Hague 1987, pgs. 33-41.

Fitzpatrick, E. A., and W. R. Stern, 1965, Components of the radiative balance of irrigated plots in a dry monsoonal environment, J. Applied Meteorology **4**:649-660.

Fleagle, R.G., and J.A. Businger, 1980, An Introduction to Atmospheric Physics, 2nd Ed. Academic Press, New York.

Fox, R.W., and A.T. McDonald, 1985, Introduction to Fluid Mechanics, 3rd Ed. John Wiley & Sons, New York.

Frevert, D.K., R.W. Hill and B.C. Braaten, 1983, Estimation of FAO evapotranspiration coefficients, J. Irrigation and Drainage ASCE **109**(IR2):265-270.

Frohlich, C., 1982, Observations of the solar constant and its variations--a summary, in: A Collection of Extended Abstracts Presented at the Symposium on the Solar Constant and the Distribution of Solar Irradiance, Boulder, CO.

Gangopadhyaya, M., 1965, Evaporation--its measurement and estimation, Int. Assoc. Sci. Hyd. Public. **68**:507-533.

Garg, H.P., 1982, Treatise on Solar Energy, John Wiley & Sons, New York, 587 pgs.

Garrity, D.P., D.G. Watts, C.V. Sullivan and J.R. Gilley, 1982a, Moisture deficits and grain sorghum performance: Effect of genotype and limited irrigation strategy, Agronomy J. **74**:808-814.

Garrity, D.P., D.G. Watts, C.V. Sullivan and J.R. Gilley, 1982b, Moisture deficits and grain sorghum performance: Evapotranspiration-yield relationships, Agronomy J. **74**:815-820.

Gates, D. M., 1962, Energy Exchange in the Biosphere. Harper & Row Publishing, New York, NY., 151 pgs.

Gerber, J.F., and J.D. Martsolf, 1979, Sprinkling for frost and cold protection, in: Modification of the Aerial Environment of Plants, Monograph No. 2, ASAE, St. Joseph, MI., 538 pgs.

Grando, F.C., 1973, Methodology existing for estimating free surface water evaporation, Int. Assoc. Sci. Hyd. Public. **108**:59.

Griffin, R.E., and E. A. Richardson, 1979, Sprinklers for micro-climate cooling of bud development, in: Modification of the Aerial Environment of Plants, Monograph No. 2, ASAE, St. Joseph, MI., 538 pgs.

Hansen, V.E., O.W. Israelsen and G.E. Stringham, 1980, Irrigation Principles and Practices, 4th Ed. John Wiley & Sons, 417 pgs.

Hargreaves, G.H., and Z.A. Samani, 1982, Estimating potential evapotranspiration, Tech. Note, J. Irrigation and Drainage, ASCE **108**(3):225-230.

Harrison, L.P., 1963, Fundamental concepts and definitions relating to humidity, in: Humidity and Moisture, A. Wexler. ed., Reinhold Publishling Co., New York, pgs. 3-80.

Hatfield, J. R., and M. Fuchs, 1990, Evaporation models, in: Management of Farm Irrigation Systems, Monograph No. 9, Hoffman, Howell, Solomon, eds., ASAE, St. Joseph, MI.

Hay, Robert K. M., and Andrew J. Walker, 1989, An Introduction to the Physiology of Crop Yield, Longman Scientific & Technical, England, pgs 17-21, 188-189, and 198.

Heermann, D. F., G. J. Harrington, and K. M. Stahl, 1985, Empirical Estimation of Daily Clear Sky Solar Radiation, Journal of Climate and Applied Meteorology, **24**(3):206-214.

Hershfield, D.M., 1964, Effective rainfall and irrigation water requirements, J. Irrigation and Drainage, ASCE **90**(2):33-47.

Hess, S.L., 1959, Introduction to Theoretical Meteorology, Holt, Rinehart and Winston, New York, 362 pgs.

Hill, R.W., C.E. Brockway, R.D. Burman, L.N. Allen and C.W Robinson, 1989, Duty of Water Under the Bear River Compact: Field Verification of Empirical Methods for Estimating Depletion, Utah Agric. Exp. Sta. Res. Rep. 125.

Hillel, D., 1980, Fundamentals of Soil Physics, Academic Press, New York, 413 pgs.

Hoffman, G.J., 1985, Evapotranspiration in saline soils, Proc. Advances in Evapotranspiration Symposium, ASAE, Chicago IL. December.

Hounam, C.E., 1973, Comparison between pan and lake evaporation, World Meteorological Organization, Tech. Note No. 126. Report as Rapporteur on Lake Evaporation of the Commission for Hydrology, 52 pgs.

Howe, Orlando, 1951, When applying irrigation water timing is important, Nebraska Experiment Station Quarterly, (printed material is in hand but exact date of publication is difficult to determine).

Howe, O.W., and H.F. Rhoades, 1962, Irrigation of Great Northern field beans in western Nebraska, University of Nebraska Experiment Station, SB **459**:1-29.

Howell T.A., 1990, Relationships between crop production and transpiration, evapotranspiration, and irrigation, in: Irrigation of Agricultural Crops, Monograph No. 30, Stewart and Nielson, eds., Amer. Soc. of Agronomy, Madison, WI.

Howell T.A., D.S. Stevenson, F.A. Aljibury, H. M. Gitlin, I.P. Wu, A.W. Warrick and P.A.C. Ratts, 1980, Design and operation of trickle (DRIP) systems, in: Design and Operation of Farm Irrigation Systems, M.E. Jensen, ed., ASAE, St. Joseph, MI.

Howell T.A., R.H. Cuenca and K.H. Solomon, 1990, Crop yield response, in: Management of Farm Irrigation Systems, Monograph No. 9, Hoffman, Howell and Solomon, eds., ASAE, St. Joseph, MI.

Hughes, William C., 1972, Simulation of salt cedar evapo-transpiration, J. Irrigation and Drainage, ASCE **IR4**:533-542.

Idso, S.B., and R.D. Jackson, 1969, Thermal radiation from the atmosphere, J. Geophysical Research **74**:5397-5403.

International Commission on Irrigation and Drainage, 1985, On updated evapotranspiration and crop water requirement terms for the ICID Multilingual Technical Dictionary on Irrigation and Drainage, 1967, ICID Bulletin **34**(1):65-66.

Iqbal, M., 1983, An Introduction to Solar Radiation, Academic Press, London, New York, Toronto, 390 pgs.

Jackson, R.D., S.B Idso and R.J Reginato, 1976, Calculation of evaporation rates during transition from energy-limiting to soil-limiting phases using albedo data, Water Resource Research **12**:23-26.

Jacobs, A.F.G., and J.H. van Boxel, 1991, Comments on: Wind profile: estimation of displacement height and aerodynamic roughness, by Abtew, Gregory and Borrelli, Technical Notes Trans. ASAE **34**(1):177-179.

Jacquot, R.G., and W.C. Porter, 1989, Interactive personal computer tools for teaching Fourier analysis, Proc. ASEE Annual Conference, Lincoln, NE., June 25-29, pgs. 628-630.

James, L.G., 1988, Principles of Farm Irrigation System Design, John Wiley and Sons, New York, 543 pgs.

Jensen, M.E., ed., 1974, Consumptive Use of Water and Irrigation Water Requirements, Report Tech. Comm. on Irrigation Water Requirements, Irrigation and Drainage, ASCE, 227 pgs.

Jensen, M.E., and H. R. Haise, 1963, Estimating evapotranspiration from solar radiation, J. Irrigation and Drainage, ASCE **89**:15-41.

Jensen, M.E., D.C.N. Robb and C.E. Franzoy, 1970, Scheduling irrigations using climate-crop-soil data, J. Irrigation and Drainage, ASCE **96**:25-38.

Jensen, M.E., J.L. Wright and B.J. Pratt, 1971, Estimating soil moisture depletion from climate, crop and soil data, Trans. ASAE **14**:954-959.

Jensen, M.E., R.D. Burman and R.G. Allen, eds., 1990, Evapotranspiration and Irrigation Water Requirements, Manual of Practice No. 70, ASCE, New York.

Jones, F.E., 1992, Evaporation of Water With Emphasis on Applications Measurements, Lewis Publishers, Inc., Chelsea, MI., 188 pgs.

Keenan, J.H., F.G. Keyes, P.J. Hill and J.G. Moore, 1979, Steam Tables, John Wiley and Sons, New York.

Kimball, H.H., 1928, Amount of solar radiation that reaches the surface of the earth on the land and on the sea, and the method by which it is measured, Monthly Weather Review **56**:393-399.

Kincaid, D.C., and D.F. Heermann, 1974, Scheduling Irrigations Using a Programmable Calculator, USDA ARS-NC-12

Kneebone, W.R., 1979, Water use by grasses, Proc. Arizona Turfgrass Conference, Tucson, AR., pgs 5-8.

Knisel, W.G., ed, 1980, CREAMS. A Field-Scale Model for Chemicals, Runoff, and Erosion From Agricultural Management Systems, Conservation Res. Report No. 26, USDA.

Kohler, M.A., and L. H. Parmele, 1967, Generalized estimates of free-water surface evaporation, Water Resource Research **3**(4):997-1005.

Kohler, M.A., T.J. Nordenson and W.E. Fox, 1955, Evaporation from pans and lakes, U.S. Dept. of Commerce, Weather Bureau Research Paper 38, 21 pgs.

Kruse, E.G., and C.M.U. Neale, 1985, Sources of error in hydraulic weighing lysimeter data, Proc. National Conference on Advances in Evapotranspiration, ASAE, Chicago, IL., December 16-17, pgs. 70-78.

Lamoreux, W.W., 1962, Modern evaporation formula adapted to computer use, Monthly Weather Review **90**:26-28.

Lewis, L. E., 1978, Development of an Evaporation Map of Wyoming for Purposes of Estimating Evaporation and Evapotranspiration, M. S. Thesis, Department of Civil Engineering, University of Wyoming, Laramie.

Linacre, E.T., 1967a, Climate and evaporation from crops, J. Irrigation and Drainage, ASCE **93**:61-79.

Linacre, E.T., 1967b, Estimation of net radiation flux, Agric. Meteorology **18**:49-69.

Linacre, E.T., 1968, Estimating net radiation flux, Agric. Meteorology **5**:49-63.

Linden, Eugene, 1990, The last drops, Environment Feature Article, in: Time Magazine, August 20, New York.

Linsley, R.K., M.A. Kohler and J.L.H. Paulhus, 1958, Hydrology for Engineers, McGraw-Hill Book Co., New York, 340 pgs.

List, R.J., 1984, Smithsonian Meteorological Tables, 5th Reprint, Smithsonian Press, Washington, DC., 527 pgs.

Makkink, G.F., 1957, Testing the Penman formula by means of lysimeters, J. Inst. Water Engineering **11**(3):277-288.

Mahrer, V., and G. Rytwo, 1991, Modelling and measuring evapotranspiration in a daily drip irrigated cotton field, Irrig. Sci. **12**:13-20.

McIlroy, I.C., and D.E. Angus, 1964, Grass, water and soil evaporation at Aspendale, Agricultural Meteorology **1**(3):201-224.

Merriam, J.L., M.N. Shearer and C.M. Burt, 1980, Evaluating irrigation systems and practices, in: Design and Operation of Farm Irrigation Systems, M.E. Jensen, ed., Monograph ASAE, St. Joseph, MI.

Merva, G.E., 1975, Physio-engineering Principles, AVI Publishing Company, Westport CO, 353 pgs.

Merva, G.E., and C. Vandenbrink, 1979, Physical properties involved in alleviating heat stress, in: Modification of the Aerial Environment of Plants, Monograph No. 2, ASAE, St. Joseph, MI., 538 pgs.

Monteith, J.L., 1965, Evaporation and environment, in: The State and Movement of Water in Living Organisms, XIX Symposium, Soc. Exp. Biol., Cambridge University Press.

Monteith, J.L., 1981, Evaporation and surface temperature, Quart. J. Royal Meteorological Soc. **107**:1-27.

Monteith, J.L., and G. Szeicz, 1961, The radiation balance of bare soil and vegetation, Quart. J. Royal Meteorology Soc. **87**:496-507.

Monteith, J.L., and M.H. Unsworth, 1990, Principles of Environmental Physics, 2nd Ed., Edward Arnold, London, 291 pgs.

Murray, F.W., 1967, On the computation of saturation vapor pressure, J. Applied Meteorology **6**(1):203-204.

Neibling, W. H., J. K. Koelliker, F. E. Ohmes, 1977, A Continuous Water Budget Model for Western Kansas, Presentation at 1977 Annual Meeting, ASAE, North Carolina State University, Raleigh, NC, Paper No. 77-2056.

Nieber, J.L., and A.S. Patwardhan, 1989, Estimation of Effective Rainfall, Research Project Completion Report, University of Minnesota, St. Paul, MN.

Nkemdirim, Lawrence C., 1976, Crop development and water loss - a case study over a potato crop, Agricultural Meteorology, **16**:371-388.

Pair, C.H., W.H. Hina, K.R. Frost, R. E. Sneed and T. J. Schlita, 1983, Irrigation, The Irrigation Association, Silver Springs, MD., 686 pgs.

Patwardhan, A.S., J. L. Nieber and E.L. Johns, 1990, Effective rainfall estimations, J. Irrigation and Drainage, ASCE **116**(2):182-193.

Penman, H.L., 1948, Natural evaporation from open water, bare soil and grass, Proc. Royal Soc. London **A193**:120-146.

Penman, H.L., 1963, Vegetation and Hydrology, Tech. Publication No. 53. Commonwealth Bureau of Soils, Harpenden, England, 125 pgs.

Peterson, G.A., T.L. Loudon and G.E. Merva, 1985, A Comparison of Evapotranspiration by Portable Chamber with Lysimeter Data, Proc. Advances in Evapotranspiration Symposium, Chicago, IL., December.

Philip, J.R., 1957, Evaporation and moisture and heat fields in the soil, J. Meteorology **14**:354-366.

Pielke, R.A., 1984, Mesoscale Meteorological Modeling, Academic Press, Orlando, FL., 611 pgs.

Pochop, Larry, and Robert Burman, 1987, Development of Evapotranspiration Crop Coefficients, Climatological Data, and Evapotranspiration Models for the Upper Green River, Project Completion Report, Project 59-6986-13, Wyoming Water Research Center, Laramie, WY., 105 pgs.

Pochop, L.O., S.A. Morton and R.L. Cornea, 1973, Psychrometric data, patterns and prediction models, Univ. of Wyoming, Water Resources Series No. 38, 20 pgs.

Pochop, L.O., J. Borrelli and R. Burman, 1984, Elevation--a bias error in SCS Blaney-Criddle ET estimates, Trans. ASAE **27**(1):125-128.

Pochop, L.O., F.M. Smith and R.E. Smith, 1985, Evapotranspiration estimates for the Pawnee Grasslands, Proc. Advances in Evapotranspiration Symposium, ASAE, Chicago, IL., December.

Priestley, C.H.B., and R.J. Taylor, 1972, On the assessment of surface heat flux and evaporation using large scale parameters, Monthly Weather Review **100**:81-92.

Pruitt, W. O., D. L. Morgan, and F. J. Lourence, 1973, Momentum and mass transfer in the surface boundary layer, Quart. J. Roy. Meteorological Soc. **99**:370-386.

Raja, V., and K.C. Bishnoi, 1990, Evapotranspiration, water use efficiency, moisture extraction pattern and plant water relations of rape genotypes in relation to root development under varying irrigation schedules, Expl. Agric. **26**:227-233.

Ratliff, L.F., J.T. Ritchie and D.K. Cassel, 1983, Field measured limits of soil water availability limits as related to laboratory-measured properties, Soil Sci. Soc. of Amer. J. **47**:770-775.

Rawls, W.J., D.L. Brakensiek and K.E. Saxton, 1982, Estimation of soil water properties, Trans. ASAE pgs.1316-1320, 1328.

Reddy, S.J., 1971, An empirical method for the estimation of total solar radiation, Solar Energy **13**(2):289-290.

Rex, D.F., 1969, Climate of the Free Atmosphere, Elsevier Publishing Co., Amsterdam, London, New York.

Rhoades, H.F., and L.B. Nelson, 1955, Growing 100-bushel corn with irrigation, in: Water, Yearbook of Agriculture, USDA.

Rijtema, P. E., and A. Aboukhaled, 1975, Crop water use, in: Research on Crop Water Use, Salt-affected Soils and Drainage in Arab Republic of Egypt. Food and Agriculture Organization Regional Office, Cairo, 92 pgs.

Ritchie, J.T., 1972, Model for predicting evaporation from a row crop with incomplete cover. Water Resource Research **8**(1)1204-1213.

Ritchie, J.T., 1974, Evaluating irrigation needs for southeastern U.S.A., Proc. Irrigation and Drainage Spec. Conf. ASCE pgs. 262-273.

Ritchie, J.T., 1985, Evapotranspiration empiricisms for minimizing risk in rainfed agriculture, in: Advances in Evapotranspiration, Proc. National Conference on Advances in Evapotranspiration, ASAE, **19**(6):1098-1103.

Ritchie, J.T., and E. Burnett, 1971, Dryland evaporative flux in a subhumid climate: II. Plant influences, Agronomy J. **63**:56-62.

Robb, G. A., T. S. Slavin, W. R. Johnston and B. C. Steinert, 1984, Crop curve development from production agriculture, Water Today and Tomorrow, Proc. Irrigation and Drainage Spec. Conf. ASCE, Flagstaff, AZ., July 24-26, pgs. 269-276.

Rosenberg, N.J., 1979, Windbreaks for reducing moisture stress, in: Modification of the Aerial Environment of Plants, Monograph No. 2, ASAE, St. Joseph, MI., 538 pgs.

Rosenberg, N.J., B.L. Blad and S.B. Verma, 1983, Microclimate--the Biological Environment, 2nd Ed., John Wiley & Sons, New York, 495 pgs.

Rumney, G.R., 1968, Climatology and the World's Climate, MacMillan Company, London.

Russell, G., 1980, Crop evaporation, surface resistance and soil water status, Agric. Meteorology **21**:213-226.

Saxton, K.E., H.P. Johnson and R.H Shaw, 1974, Modeling evapotranspiration and soil moisture, Trans. ASAE **17**(4):673-677.

Schneckloth, J.P., N.L. Klocke, G.W. Hergert, D.L. Martin and R.T. Clark, 1991, Crop rotations with full and limited irrigation and dryland management, Trans. ASAE **91**:2372-2380.

Sellers, W.D., 1965, Physical Climatology, Univ. of Chicago Press, Chicago, IL., 272 pgs.

Sharma, D.K., A. Kumar and K.N. Singh, 1990, Effect of irrigation scheduling on growth yield and evapotranspiration of wheat in sodic soils, Agric. Water Mgt. **18**:267-276.

Siemer, E.G., and C. B. Rumberg, 1975, Management practices determine plant species in meadows, Colo. State Univ. Agric. Exp. Sta. Bull. 564-S, 12 pgs.

Silver, H.F., and J.E. Nydahl, 1977, Engineering Thermodynamics, West Publishing Co. New York, 557 pgs.

Singh, B., and G. Szeicz, 1978, Predicting the canopy resistance of a mixed hardwood forest, Agricultural Meteorology, **21**:49-58.

Spencer, J.W., 1971, Fourier series representation of the position of the sun, Search **2**(5):172.

Stamm, G.G., 1967, Problems and procedures in determining water supply requirements for irrigation projects, Chapter 40 in: Irrigation of Agricultural Lands, by Hagan et al. Amer. Soc. of Agron. II, Madison, WI.

Steiner, J.L., 1989, Tillage and surface residue effects on evaporation from soils, Soil Sci. Soc. of Amer. J. **53**:911-916.

Stewart, J.I., and R.M. Hagan, 1973, Functions to predict effects of crop water deficits, J. Irrigation and Drainage, ASCE **99**IR4:421-439.

Stewart, R.B., and W.R. Rouse, 1976, A simple method for determining the evaporation from shallow lakes and ponds, Water Resources Research **12**:623-628.

Stewart, J.I., R.M. Hagan and W.O. Pruitt, 1974, Functions to predict optimal irrigation programs, J. Irrigation and Drainage, ASCE **100**IR2:179-199.

Stine, William B., and Raymond W. Harrigan, 1985, Solar Energy Fundamentals and Dsign with Computer Aplplications, John Wiley & Sons, New York, NY, 536 pgs.

Sweers, H.E., 1976, A monograph to estimate the heat exchange coefficients at the air-water interface as a function of wind speed and temperature: A Critical Survey of Some Literature, J. of Hydrology **30**:375-401.

Szeicz, G., and I. F. Long, 1969, Surface resistance of crop canopies, in: Water Resource Res. **5**:622-633.

Szeicz, G., C.H.M. Van Bavel and S. Takami, 1973, Stomatal factor in the water use and dry matter production by sorghum, Agric. Meteorology **12**:361-389.

Tetens, O., 1930, Uber einge meteorplogische Begriffe, Z. Geophys. **6**:297-309.

Thom, A.S., and H.R. Oliver, 1977, On Penman's equation for estimating regional evaporation, Quart. J. Royal Meteorological Soc. **103**:345-347.

Thompson, N., I.A. Barrie and M. Ayles, 1981, The meteorological office rainfall and evaporation calculation system: MORECS (July 1981), Hydrological Memorandum No. 45, Meteorological Service, United Kingdom.

Thorndike, E.L., and C.L. Barnhart, 1974, Thorndike Barnhart Advanced Dictionary, 2nd Ed. Scott, Foresman and Company, Dallas, TX., 1186 pgs.

Thornthwaite, C.W., 1948, An approach toward a rational classification of climate, Geograph. Rev. pgs. 38,55.

United States Department of Agriculture, 1970, Irrigation Water Requirements, Soil Conservation Service Tech. Release No. 21, (rev.) 92 pgs.

United States Department of Commerce, 1972, Substation Observations, National Weather Service Observing Handbook No. 2.

United States Department of Commerce, 1982a, Evaporation Atlas for the Contiguous 48 United States, NOAA Technical Report National Weather Service 33.

United States Department of Commerce, 1982b, Mean Monthly, Seasonal and Annual Pan Evaporation for the United States, NOAA Technical Report National Weather Service 34.

United States Department of the Interior, Geological Survey, 1981, National Handbook of Recommended Methods for Water-Data Acquisition, U. S. Government Printing Office, Washington, DC.

United States Naval Observatory, 1946, Tables of Sunrise, Sunset and Twilight (Supplement to the American Ephemeris), Washington, DC., 196 pgs.

United States Naval Observatory, 1988a, The Astronomical Almanac for the Year 1989, joint publication of the United States Naval Observatory, the Nautical Almanac Office and Her Majesty's Nautical Almanac Office, Royal Greenwich Observatory, U.S. Government Printing Office, Washington, DC., and Her Majesty's Stationary Office, London.

United States Naval Observatory, 1988b, Astronomical Data for the Year 1990, joint publication of the United States Naval Observatory, the Nautical Almanac Office and Her Majesty's Nautical Almanac Office, Royal Greenwich Observatory, U.S. Government Printing Office, Washington, DC., and Her Majesty's Stationary Office, London.

Van Wijk, W.R., 1963, Physics of Plant Environnment, W.R. Van Wijk, ed., North Holland Publishing Company, Amsterdam, 382 pgs.

Van Wylen, G.J., and R.E. Sonntag, 1985, Fundamentals of Classical Thermodynamics, 3rd Ed. John Wiley & Sons, New York, 722 pgs.

Vaux, H.J. Jr., and W.O. Pruitt, 1983, Crop-water production functions, in: Advances In Irrigation, D. Hillel, ed., Academic Press.

Wallace, C.H., C.H. Batchelor and M.G. Hodnett, 1981, Crop evaporation and surface conduction calculated using soil moisture data from central India, Agric. Meteorology **25**:83-96.

Walter, I.A., E.G. Siemer, J.P. Quinlan and R.D. Burman, 1990, Evapotranspiration and agronomic responses in formerly irrigated mountain meadows, South Park, Colorado, Report for the Board of Water Commissioners, City and County of Denver, CO., March 1, 1990, 204 pgs. plus appendices.

Wark, K., 1983, Thermodynamics, McGraw-Hill Book Company, 895 pgs.

Watts, D.G., and W.L. Kranz, 1982, Improvement of irrigation scheduling techniques for corn with variable crop maturity range, plant population and water supply availability, Termination Report, OWRT Project No. A-057, Dept. of Agric. Engr. University of Nebraska, Lincoln.

Weiss, A., 1977, Algorithms for the calculation of moist air properties on a hand calculator, Trans. ASAE **20**(6):1133-1136.

Weiss, A., 1982, An experimental study of net radiation, its components and prediction, Agronomy J. **74**:871-874.

Welty, J.R., C.E. Wicks and R.E. Wilson, 1976, Fundamentals of Momentum, Heat, and Mass Transfer, 2nd Ed., John Wiley & Sons, New York, 789 pgs.

Williams, J.R., A.D. Nicks and J.G. Arnold, 1985, Simulator for water resources in rural basins, J. Hyd. Engr., ASCE **111**:970-986.

Woolf, H. M., 1968, On the computation of solar evaluation angles and the determination of sunrise and sunset times, National Aeronautics and Space Administration Report NASA TM-X-164, September.

Wright, J.L., 1979, Recent developments in determining crop coefficient values, Proc. Irrigation and Drainage Spec. Conf. ASCE, Albuquerque, NM., July 19.

Wright, J.L., 1981, Crop coefficients for estimates of daily evapotranspiration, Symposium Irrigation Scheduling for Water and Energy Conservation in the 80's, ASAE, Chicago, IL., December 18-26.

Wright, J.L., 1982, New evapotranspiration crop coefficients, J. Irrigation and Drainage, ASCE **108**(IR2):57-74.

Wright, J. L., and M. E. Jensen, 1972, Peak water requirements of crops in Southern Idaho, J. Irrig. & Dr. Div. ASCE, **96**(IR1):193-201.

Index

GLOSSARY

evaporation-is the process where a liquid is transformed entirely into a gas. In this volume water is the liquid and water vapor is the gas. Evaporation is abbreviated as E.

evapotranspiration-the combined process of both evaporation from soil surfaces and transpiration through plant surfaces. The term Consumptive Use is the same as ET with the exception that it includes water needed for the building of plant tissue. The term consumptive use originated in the Western United States and may be considered to be somewhat archaic.

Joule-a unit of work equal to a Newton-meter in the S.I. system.

management allowed deficiency- the percent of the available soil moisture in the root zone that corresponds to a management allowed crop stress.

pan evaporation-is evaporation from a vessel constructed to contain water and which is placed so that evaporation occurs caused by the local environment near the pan. E_p has been used as an index of evaporation for reservoirs or ET. Many different kinds of pans have been constructed. Variations include different sizes, materials of construction including color and whether screened or not. The type of evaporation pan and exposure of the pan must be carefully considered when utilizing E_p data.

plant (crop) coefficient-a dimensionless ratio used to relate ET_p or ET_r to ET from a specific crop or plant at a specific time. See sections YYY for more information.

plant (crop) ET-Is the ET from a specific plant grown under defined conditions including soil moisture, fertility and other cultural conditions. ET_a has been designated as actual or crop ET.

potential ET-the rate at which water would be removed from moist soil if available.

potential evapotranspiration-has been used in a variety of ways. It always has referred to plants adequately supplied with moisture and usually not limited by disease or fertility.

reference plant (crop)-The same thing as ET_p except it is applied to an identifiable crop such as alfalfa or grass. ET_r is ET from a well adapted local crop grown under the same conditions for ET_p. See sections ZZZ for further definition.

reference plant (crop) evapotranspiration-is an outgrowth of the confusion surrounding ET_p. It is used as a standard ET used for the prediction of ET by using crop coefficients. ET_r is ET from a well adapted local crop under the general conditions of ET_p.

saturation vapor pressure-the equilibrium vapor pressure where molecules leaving the water surface is equal to the molecules entering the water surface. Saturation vapor pressure is a function of temperature only.

transpiration-is the process where liquid water within a plant is transformed into water vapor in the air surrounding the plant.

vapor pressure-the partial pressure caused by water molecules in the atmosphere.

Watt-a rate of doing work equal to a Joule sec^{-1} in the S.I, system.

wet bulb temperature-technically the temperature a parcel of air would have if cooled adiabatically to saturation, the temperature shown by a thermometer covered by a wet gauze sleeve in a psychrometer. Psychrometers are ventilated, non ventilated and mechanically aspirated. Wet bulb temperature from a ventilated psychrometer should be from a full size psychrometer being rotated at 3 and one half r.p.m.